T0331779

GOD: The Evidence

" Lord, I believe,

help me in my unbelief."
Mark 9:24

RONALD W. LARSEN PHD

authorHOUSE

AuthorHouse™
1663 Liberty Drive
Bloomington, IN 47403
www.authorhouse.com
Phone: 833-262-8899

Published by AuthorHouse 01/17/2023

ISBN: 978-1-7283-7010-1 (sc)
ISBN: 978-1-7283-7011-8 (hc)
ISBN: 978-1-7283-7154-2 (e)

Library of Congress Control Number: 2020916195

*On the Cover: Hubble Space Telescope image of an infant emission-line
star located 2,300 light-years from earth. Courtesy of NASA.*

Print information available on the last page.

TABLE OF CONTENTS

Dedicated to the loving
memory of my father.

"The known is finite, the unknown is infinite; intellectually, we stand on an islet in the midst of an illimitable ocean of inexplicability. Our business in every generation is to reclaim a little more land." Thomas H Huxley

ACKNOWLEDGEMENTS

I would like to thank Frank Braxton (Brack), and Gordon Shugars for reviewing the manuscript. I would like to commend the professionalism of the technical support team at Author House, I would especially like to recognize Mark Francis for his words of encouragement. Finally, I would like to thank Kristine Mayo for her contagious enthusiasm for this little project.

I would like to recognize biologist and atheist Richard Dawkins who challenged someone to write this book. I would like to recognize eminent physicist and Hebrew scholar Gerald Schroeder for his ideas that put me on this path. Finally, I would like to recognize the greatest physicist who ever lived, Albert Einstein. He is an inspiration to anyone who wants to know about the amazing world beyond our little planet. He is a guidepost at every turn in this book.

PREFACE

"The universe knew we were coming." [1] Freeman Dyson, Manhattan Project, British Royal Society, Professor Emeritus of Physics at Princeton

Science is contributing new and important insights into the nature of reality. Science and technology also reveal with increasing clarity the place of humanity in the cosmos. In the last century, modern science achieved a stature above the fray by avoiding cultural biases, religion in particular. Beliefs that do not accept the validity of science jeopardize their credibility. Science is now the gold standard for the truth about the natural world. But the methods of science will never get us to the goal line. In science, there may be no goal line, and even if there is, there is no way for science to recognize it. Physics professor Brian Greene quotes humorist Stephen Wright, "Physics is like the photographer trying to get a close-up shot of the horizon." [2] We must look beyond the horizon of science for better answers.

Any way you look at it, we are fortunate to be here. But exactly how did we get here and why are we here? Our intellect compels us to try to find answers. Our significance in the grand scheme of things cannot be denied. Our ancestors knew instinctively a higher power than themselves manifest in the cosmos. They knew instinctively we were here at the pleasure of a force far more significant than themselves. Everything that can be seen and heard, from the myriad stars to rolling thunder, is telling the greatness of an Almighty power. Whether we like it or not, this is the only logical explanation for our existence. Science and philosophy are merging into a coherent worldview, which combines beautifully with the God Hypothesis. Physicist Max Born said, "I am now convinced theoretical physics is philosophy." Science, philosophy, and religion probe the nature of our existence, but from different perspectives. While religious beliefs still prevail, they have been forced to give up a lot of turf to scientific progress. Is science capable of answering the deepest questions, thereby eliminating the need for God, religion, and philosophy? Can science and religion both be right? Can reasoned discourse explain apparent discrepancies between science and Scripture? The acrimony between science and religion heated up during the last century due to the strength of religious fundamentalism and the threat of evolution to their beliefs. Prominent scientists increasingly express their opinion that faith in God is just superstition.

While a student at Caltech, I heard Nobel physicist and atheist, Richard Feynman, handily dominate debates with theologians. I've heard preachers railing against evolution. The issues are divisive, emotionally charged, and debated vigorously. The culture clash between our religious heritage and the inexorable progress of science and technology has become a detriment to our way of life as each disparages the other. Science has undoubtedly exposed many flaws in religious beliefs. For two millennia, the incompatible worldviews of our Greco-Roman and Judeo-Christian heritage have yet to be fully resolved. But many great thinkers have pressed hard on the issue and it appears significant progress towards resolution may be possible.

10/20/2020 CNN: In the midst of the disastrous public response to the lethal covid-19 pandemic, Dr. Anthony Fauci, Head of the Center for Infectious Disease Control, remarked, "We are going through a time that is disturbingly anti-science by a certain segment of our society." He and his family received multiple death threats. What concerns me, he is referring to the right-wing fundamentalist Christian agenda.

The Catholic Church has finally consented to scientific progress. A letter from the Vatican written at the time of the two-hundredth anniversary of Charles Darwin's birth declared, "The purpose of Scripture and religious truth is not to convey scientific information, but to transform hearts. To mislead children with false information is not benign. It's irresponsible, repressive, and dangerous." The Vatican has finally taken the lessons of the Inquisition and the excoriation of Galileo to heart. But why did it take two centuries for a pope to recant a lie? Jonathan Swift famously wrote in 1710, "A lie can travel around the world while the Truth is putting on its shoes." So, walk a mile in my shoes, then see what you think.

America is first and foremost founded on a noble ideal written on a piece of paper. Thomas Jefferson put our religious values into his iconic Declaration of Independence on July 4, 1776: "all men are created equal" by their Creator. This is much more than a political declaration of humanistic democracy. He is declaring there is no inherent right of human equality in nature; equality derives from God. Jesus was the first to declare the idea of equality of men as well as women. America has brought Christian democracy to the world, promoting liberty and justice for all, opposing tyranny, oppression and autocracy, tempering the vagaries of capitalism, but we have not eradicated prejudice. In 1992, street philosopher, Rodney King, famously said, "Why can't we all just get along." Equality in practice defies common sense because every person is born to become a uniquely different individual. The cosmos is so grand it makes all men equal in comparison. Thanks to science, we now know all humans are 99.9% genetic equals. Benjamin Franklin wrote, "The worship of God is a duty." He also wrote, "Only virtuous people are capable of freedom." Thomas Payne wrote, "The mind, once enlightened, cannot again become dark." A great modern patriot, Senator John McCain, wrote, "Nothing in life is more liberating than to fight for a cause greater than yourself." As an officer of the Freemasons, George Washington stated a member "will never be a stupid atheist nor an irreligious libertine." On the reserve currency of the world, America proudly declares, "In God we trust." In 1954, "under God" was added to the Pledge of Allegiance, over strong objection by atheists. I have great respect for the integrity of my atheist friends. They have a right to be heard. I fear religion is becoming irrelevant. My motivation is to find better answers for my beliefs. Atheists tend to be discriminated against by the religious majority. An atheist president in America is almost unthinkable. Moribund religious institutions are in dire need of a challenge to wake them up to a rapidly changing world. Whether good or bad, whether society is in moral decline or not, it is clear educators and scientists who promote atheism are challenging the greatest threat to the sacred trust of our Founding Fathers. Atheism is on the rise in America for several reasons. It is my premise that a significant problem is a persistent and growing discrepancy between science and religion. Atheism is a core belief disguised as intellectualism. It is astonishing to me that biologists are at the forefront of the atheistic movement considering they have no clue how to explain the most critical scientific mystery in history, the origin of life. Oxford biologist, atheist, and iconoclast, Richard Dawkins, declared, "Religions have historically always attempted to answer questions properly belonging to science. I think a case can be made that faith is one of the world's greatest evils, comparable to smallpox, but harder to eradicate. Don't fall for the argument that religion and science operate on separate dimensions or are concerned with quite separate sorts of questions." It is ironic that an atheist considers himself an expert on evil. Thomas Aquinas would answer, "To one who has faith, no explanation is necessary; to one without faith, no explanation is possible." Biology professor and atheist Jerry Coyne explains, "In religion, faith is a virtue; in science, faith is a vice." The implication is science is sound and religion is wrong. Eradicating faith won't be easy. Hope is a response to the total human experience, heart, mind, and soul, which goes way beyond the current reach of science. In science, faith in the method is fundamental. The scientific method is only concerned with what can be observed. In religion, faith in the invisible God is fundamental. Faith in science and faith in religion are not the problem. Faith in humans, such as Coyne and myself, is the problem. We must rely on experts in their respective fields in order to shed light on the confounding issues addressed in this book.

On the other hand, scientific discovery and religion are a leap into the unknown. Humans are not mind readers. No one knows what lurks in the minds of others. We must trust the human spirit, or we may not survive. Likewise, we cannot understand the Mind of God, but we are asked to believe him. 92% of the National Academy of Sciences are atheists. A recent Pew poll found younger Americans are 31/2 times more likely to doubt the existence of God than their parents. According to a new Gallup survey, more than 3 in 4 Americans say religion is losing its influence in the U. S. The same people say this trend bodes ill for the country. On the other hand, the American public is woefully ignorant when it comes to science. In a recent Gallup poll, only 15% of the population said they believe in the theory of evolution. 64% believe the universe was created around 6 to 10 thousand years ago. The flat earth society is alive and well. We live in the era of science and technology; there is no turning back; facts like global warming matter whether we like it or not. Unfortunately, the hysteria over scientific facts like evolution, age of the universe, and global warming are driven by misguided religious politics. Freeman Dyson said, "Anyone who rejects scientific theory is free to believe whatever he wants. He may believe the cosmos exists on the back of a turtle or the moon is made of green cheese. To such people, reason and logic are irrelevant." [1] Neurologist Stuart Firestein points out, "Knowledge is a big subject; ignorance is a bigger one." Evolution is no longer a scientific theory; like gravity, it is a scientific fact. Carl Sagan observed, "We live in a society exquisitely dependent on science and technology, in which hardly anyone knows anything about science and technology. The scientific "theory of everything" will be taught in schools as though no God is needed. Parents should be prepared to explain why there must be a Creator and why teachers are expected to avoid talking about religion in class. This is the way it must be until physicists let God into their theories because all other hypotheses have been exhausted; I believe this time is coming. The scientific dilemma concerning God is analogous to a quip by a British leader, "Americans will always do the right thing, after exhausting all the alternatives" Whatever your beliefs, you need to know why the scientific theory of creation is so vital to the understanding of the creation story in Genesis.

The late professor of literature at Oxford, C. S. Lewis said, "A man who wishes to remain a sound atheist cannot be too careful of his reading." [3] In other words, God is the only foundation which justifies and explains existence. Once the necessity of a Creator is established by scientific evidence, we can be confident the Bible has not led us away from the search for Truth. Science, religion, and philosophy come together to give us the best knowledge about the true nature of reality.

I was raised a Christian, and now I am a physicist and engineer. I quickly learned it was a fatal career move to talk religion with coworkers. It didn't take long before I didn't go to church or socialize with Christians. When I retired, I started attending church again. My golfing buddy is an atheist. I felt empathy for him because he had a miserable childhood. Every time he hit a bad shot, he would say, "god damn it." Finally, I asked, "Why do you swear at God when you don't believe in him." The next time we played, one of the other golfers came up to me and asked, "Why does your friend cuss at you?" I said, "Oh my." He said, "Every time he hits a bad shot, he says, Ron damn it."

I tried to talk a little religion with him. I failed miserably. I struggled with the puzzle of what to tell my children. So, I wrote a book, "Modern Science Meets Old Time Religion." All my friend ever said was, "It rationalized what I already believed." One golfer sang "Give Me That Old Time Religion." That book went into the circular file. In this revision, religious beliefs are only discussed in the context of scientific evidence and reasoned argument. Here, widely recognized subject matter experts will express their views for and against the God Hypothesis. Who would dare challenge Albert Einstein, who said, "He, who can no longer pause to wonder and stand rapt in awe, is as good as dead; his eyes are closed?" [4]

Most people who know me will be surprised by my views on the Bible and the cosmic mystery theater. As a child, I was taught to believe in Santa Claus and Jesus. My parents knew nothing of the culture shock, which would greet me in college. They told me I could not believe in evolution and be a Christian. I wanted to know whether faith is necessary to believe in a Creator. If it is, then the politics of religion by our Founding Fathers were misguided. It turns out to be relatively straightforward to demonstrate the Creator is a valid scientific hypothesis, but a personal God is another matter altogether. While surfing the internet, I ran across an explanation for physical reality by Freeman Dyson, an eminent physicist and Christian. Most scientists are giving society a clear signal that the religious definition of Truth is outdated, harmful, and must be eradicated. Most scientists now assert there is no evidence of the supernatural nor spiritual reality. Scientists couch questions about existence in terms of pure empirical, i.e., material, evidence. So far, the scientific worldview is catching on with young people, but the response in the public square is decidedly mute and fearful of change. Science is seen as the religion of atheism. So far, faith tempered by reason is the social norm. In the meantime, scientists are stuck duking it out with the moral majority. Politicians know which side of their bread is buttered.

"I find it as difficult to understand a scientist who does not acknowledge the presence of a superior rationality behind the existence of the universe as it is to comprehend a theologian who would deny the advances of science." Werner von Braun, Director Emeritus of the NASA Rocket Research Program.

CHAPTER 1

INTRODUCTION

Brief History of Dirt

"Dust you are and to dust, you will return." Genesis 3:19

"The cosmos is within us; we are made of star stuff. We are a way for the cosmos to know itself." Carl Sagan, astronomer

Clumps of nuclear waste rose from the ashes of a star to wonder about their existence. Known as Sapiens, i.e., wise dirt, they were bipeds who believed the sun was their creator. Then the wisest dirt said, "The sun is not great dirt like us." To settle the matter, they worshiped all kinds of high-grade dirt, ancestors, animals, rulers, and idols. Then the wisest dirt said, "No! Our creator must be the "Greatest of dirt." With God, they did not even need an army. And it came to pass, belief in God was cast into tall pointy stone buildings across the land. Then dirt in lab coats said, "Dirt is just dumb atoms." The guys in lab coats broke an atom into pieces and looked inside. They declared the Higgs boson made atoms.

We now know the dirt under our feet, and the air we breathe are the same atoms from which we are made. The biosphere has recycled those same atoms for billions of years. One question remains: What would have made plain old dirt want to get a life? There is only one progenitor for all of life. This event occurred only once, and it cannot be explained by the lab coats, although they are sure they will eventually discover how life started in a pool of scum. Then they will say, "God is not needed because we know how it all began." Already, the most vocal scientists are atheists, like Stephen Hawking, who declared we are nothing more than "chemical scum." We are nothing more than a pile of atoms studying atoms. The increasingly popular nihilist view is we are staring at an empty universe devoid of meaning. But this ignores the fact that humans are creatures who evolved out of nature to acquire a godlike mind. Our dual nature is the puzzle of the human condition.

We cannot know anything that exceeds the limits of our collective minds. From the beginning, humankind knew instinctively something else must be at play. The fact intelligent and self-aware beings emerged from dirt strongly suggests a mysterious and unknowable intelligence pervades the cosmos. Humans would like to make smart and self-aware robots. All humans can do is copy and reuse what nature has already done. Does this make humans as intelligent as nature? Hardly! It can only be concluded the miracle of life is the smoking gun revealing once and for all humans are the beneficiaries of intelligent design beyond our comprehension. We have been given the keys to the cosmos. What other secrets do the cosmos hold? Humans are the cosmos trying to understand themselves. It is in our genes to want to know: Why are we here? Six-thousand years ago, a wise tribe realized their God made all living things, including the first godlike creature, Adam, from dirt. Nothing has happened to change the story.

Ever since humans picked up a rock, they have found good and bad things to do with them. Freud noted, "The first human who hurled an insult instead of a rock was the founder of civilization." But real truth remains in short supply. It is debatable whether or not science is a search for truth. Science has proved its worth many times over, but it is well known its methods somewhat limit science to observable facts. But the gods of science are the new arbiters of truth. Along the way, science has found an answer of biblical proportions; a blast from the past created the cosmos. The debate is closing in on the truth of a real kind. Both wisest dirt and science dirt believe in the big bang. The gods of science say, "Not so fast." The cosmos created itself from nothing. No God is needed. Religion is the enemy of science because it shuts the door on the search for knowledge. Faith has made a wrong turn on a one-way street. The quest for knowledge must continue without interference. Science and religion are fighting a caged deathmatch. The legacy we inherited from our ancestors is a never-ending search for truth.

Atheists know science works, but are certain religion doesn't. They are convinced reason and science can provide all the answers we need. After all, they can claim science is humankind's highest achievement. All of the human experience can be reduced to the material world. Any knowledge outside of science is just wishful thinking. Atheists believe seeking more profound answers is an exercise in delusion. They see religion offering false hope to the emotionally and mentally impaired. Humans were not made in the image of God. Man is just a collection of atoms, a creature of evolution.

Standing at the Gate

"God used beautiful mathematics in creating the world." Paul Dirac, 1933 Nobel in physics

The year was 1543 when Copernicus, an undistinguished Polish scholar, gave birth to modern science. His discovery was largely ignored because mathematics and geometry were quite complicated. A century later, Galileo brought the discovery to the world. It changed the course of history. Copernicus proved that the sun does not rise in the morning. The earth turns to face the sun each day. The rising sun is just an illusion because we cannot feel the earth turning.

Copernicus proved the earth, and the planets revolve around the sun. Man was no longer at the center of the cosmos, as written in the Holy Book. We still say the sun rises because that is what appears to happen. Why would ancient biblical writers think otherwise? It would only add unnecessary confusion to their dilemma and confound their view of the world even more. Such knowledge was irrelevant to their lives and would only detract from God's purpose for them. The secrets of God's creation would have to wait until mankind acquired the knowledge to comprehend them. That privilege has been given to us today. We have glimpsed the glory of God's creation. This is the first and foremost indication that we are little gods in our world; we are made in the image of God, *imago Dei*.

The year was 1929. An earth-shattering discovery equal to the Copernican revolution was made by a little-known astronomer Edwin Hubble. Scientists scoffed at him. His discovery was turned into an inside joke called the big bang. It was not until 1965 that the big bang was confirmed by two physicists at Bell Labs, Arno Penzias and Robert Wilson, who receive the Nobel in 1978. Penzias wrote, "When we look at the universe and its vastness, then life on our earth, how can one not be amazed at all we see? How can one think that all that would be needed, could just happen? Even if so, what put it all into motion, and where did those things come into being? We don't see things like that just happening in science. Ruling things out in advance is actually bad science. Inserting wanted, and held beliefs and worldviews, and making the science "behave" as it were, doesn't work. That some are not wanting to even entertain the kind of intelligent being needed to make something even begin to happen, (since we know there was a beginning...), is something that is true of many people, but the putting up of intellectual smokescreens is more desirable. The truth, I think is better for all mankind in the long run anyway, even if it seems undesirable at the moment. It can be the beginning of a lifelong, amazing search to really discover our roots. No need to be afraid."

The big bang affirms that the cosmos had a beginning, just as foretold in Genesis. Even more astounding, Einstein's laws of relativity require that the big bang begins from nothing. The Pope declared the big bang proved the existence of God. He recanted when very vocal scientists vehemently declared the Pope was wrong to meddle in science.

Here we encounter the biggest deception ever perpetrated upon a gullible public. World-renowned physicists totally reject this discovery as evidence for God. Instead, they claim just the opposite, that this discovery proves there is no need for God. Scientists don't have a clue where the laws of the cosmos came from. Elite scientists who are atheists want us to believe they invented them, rather than discovered them. They ignore the fact that Genesis tells us that God is the Great Lawgiver. He spoke the laws by which we should live and He spoke the laws which created and sustain the cosmos. God creates from nothing, *creatio ex nihilo*.

God's first act was to create light. Light is the Great Informant of the cosmos. Without light, we would not just be lost in darkness; we could not exist. Light sets the speed limit of the cosmos. The fact light seems instantaneous is just an illusion because we are so small in a grandiose cosmos. It takes eight minutes for light from the sun to reach earth. Light allows us to see inside stars and see the remnants of the big bang. Einstein proved gravity obeys the same rules as light, including the speed of light. If the sun suddenly disappeared, it would take eight minutes for the earth to know it was supposed to stop revolving around where the sun had been.

To physicists, light is more than the preeminent source of energy; it is the preeminent source of information. Physicists have discovered that information is more fundamental to existence than matter, energy, space, or time. As you will see, the cosmos is a magnificent information system. Our brains happen to be sophisticated information systems, too. Physicists know there is a cosmic source of information that drives the cosmos. British physicist, astronomer, and mathematician Sir James Jean observed, "The universe seems to me to be nearer to a great thought than to a great machine."

Something out there is communicating with us, and I don't mean space aliens. Cosmic information was sent out into a void at the instant of creation. This is indisputable evidence of the mind of God. Physics shows that the cosmic information system is the bedrock of physical reality. God is the objective reality that created and sustains material reality. Physics has demonstrated that material reality is just an illusion. Science has given us intellectual insight into the metaphysical nature of objective reality. Objective reality is the eternal presence of God, which is experienced through spiritual awareness. Humans can free their minds from the confines of the finite illusion and see themselves in communion with the eternal Source of existence.

The conscious mind can literally learn to tune in or tune out of the illusion to see through the veil which separates us from God.

Don't think for a moment that there is a place for mindless accidents in science or the cosmos. A scientific accident is a dropped test tube. Flipping a coin is not the cause of existence. Like everything else, probability in nature is governed by immutable laws. Causality is always present, even in the face of apparent chaos.

A century ago, Congress proposed closing the Patent Office because everything had been invented. Since then, science and technology have erupted into a frenzy of discovery and innovation that has changed the way we live and work. Physicists cannot patent a cosmos they didn't invent. All they are doing is reverse engineering what already exists. So, the question remains, who or what did all these things we know, and why were they done? The only way to give meaning to God is through science and Scripture. In the Holy Book, God is described as unseen, eternal, immutable, absolute, omnipresent, and omniscient. As you will see, these attributes are also foundational principles of the laws of physics. If the laws of the cosmos did not have these properties, there would be no cosmos.

Search for Truth

"Come now, let us reason together, says the Lord." Isaiah 1:18

Epistemology is the philosophy of knowledge. Plato's definition of knowledge as justified true belief is illustrated in the Venn diagram. Belief is the basis for individual thought, whereas knowledge is acquired by reasoning about objective truth, which is derived from indisputable facts. The definition presupposes human knowledge is inherently imperfect because we can't always be sure we know all the objective truth.

Plato's Definition of Knowledge

Science asks "how" questions from the bottom up while religion asks "why" questions from the top down. In the case being considered, science is concerned with how we got here while the Bible is concerned with why we are here. Science qualifies as knowledge, but religion does not. Religious belief is a matter of personal either faith or opinion, rather than knowledge. Theologians say they know when they don't know. Faith needs to be fortified by objective truth to fend off doubt. This does not mean rational thinking is superior to intuitive thinking. It means arguments from Scripture or ignorance risk ridicule.

A Venn diagram of a religion is a small circle contained wholly within a larger circle. The small circle represents knowledge derived from facts, what is known as objective truth. The larger circle represents revealed truth unique to a particle religion. For example, the Bible is the revealed truth of the Jewish culture. What is revealed, truth is not objective truth. It is the cultural lore and legends of a given culture. Religious faith derives from personal experience, which may not be very convincing to others. So, this book is an examination of the objective truth that makes the Bible the greatest religious document ever written.

God is not an empirical claim, so his existence cannot be proved nor disproved directly from empirical evidence. However, as will be argued, there is sufficient circumstantial scientific evidence to infer the existence of God beyond a reasonable doubt. The God Hypothesis is a scientifically based theory of creation called the Creator's Voice. If science can give us an answer to the God question, it will inoculate the believer from doubt. Even if you are a skeptic, you should want to know scientific and philosophical knowledge both for and against the God Hypothesis.

Physics is the most fundamental of all sciences. A physics theory is an objective truth because it must be correct with 99% confidence. All other sciences must be explained in the terms of physics to claim this level of confidence. Never argue against a theory of physics; you will be wrong. Physicist Ernest Rutherford said, "All science is either physics or stamp collecting." For example, biologists cannot explain how life began. Their current hypothesis is that life dropped out of the sky.

In modern physics, knowledge and truth exist, just not in all the ways we thought. Modern physics is undergoing an earth-shattering revolution. Classical physics assumed empirical facts are independent of the observer. Modern physics is confounded by the fact this is not true. Atoms are no longer the most fundamental building block of matter. The elementary particles of matter are made from fields of information, just like light and gravity. The universe only exists as an illusion created by the mind of God. The Creator's Voice postulates the universe is a communication pathway between the mind of God and the minds of mankind.

The first chapter of Genesis is a poetic explanation for the creation of the universe. The Bible rests on the small shoulders of this ancient story. The story need not be literally true to understand its purpose. Fundamentalists believe the story is the spoken word of God. If the story is incompatible with scientific facts, it is properly considered a metaphor or myth. For this reason, a rational examination of the whole Bible is in order.

The Bible did not just drop from the sky. There is no logical reason why it would agree with modern science. The Jewish people were consumed by religious customs. Neighboring cultures were far more sophisticated in many scholastic disciplines, including mathematics, geometry, astronomy, and natural philosophy. The Israelites were aware of some of this knowledge. The Greeks believed the earth sat at the center of the universe. This erroneous "fact" was so obvious everyone believed it, including the Jews. This falsehood appears in the Bible. The scientific revolution began in 1543 when Copernicus proved the earth rotated around the sun. This one scientific fact drove a wedge between science and Christianity for four centuries. Today, evolution is the wedge issue.

The premise here is that science and the Bible are telling the same story, from different perspectives. The Bible is the Creator's story, and the Creator has given science the privilege of reading his creation. Just as our brain is designed to give us two very different views of reality, so it is that the Bible is intuitive and science is analytical. Both forms of knowledge are equally valid. Together, they reveal knowledge of the deepest secrets of reality. Together, they offer a comprehensive and complete worldview to lead mankind into the future. The modern world has left the dark ages of the past as we move into the light of the scientific adventure, which is opening the secrets of reality as never before. The next time someone tells you something is correct, but you are not sure about it, ask them if they know it because of some evidence.

Basic Facts

'Everyone is entitled to their own opinion, but not their own facts," Senator Patrick Moynihan

The history of Western Civilization began with the cultures of secular Greece and the religious Jews. In order to analyze the Bible using today's knowledge, the state of knowledge 2,000 years ago provides the context from which to begin. Here is a summary of the relevant "facts" as they were believed to be true at that time:

- Aristotle believed the universe was infinite. The Jews believed the universe was finite.
- The Greeks believed three gods, Chaos, Cronos, and Uranus sprang into existence from a void. The Jews believed in one eternal God.
- The Greeks believed three dozen gods were born to make the universe. The Jews believed one and only one God created the universe from nothing but his spoken words.
- The Greeks believed their gods ruled the universe. The Jews believed God's spoken laws govern the universe.
- The Greeks believed Apollo was the sun god. The Jews believed the sun was a lamp.
- The Greeks believed matter is made of atoms. The Jews believed all things were made from God's spoken words.
- The Greeks did not specify a creation time table. The Jews specified an orderly creation process over six time periods.

Science proves the universe is intelligible. Intelligent information cannot create itself. Intelligent information, which created and sustains the cosmos, is contingent on an intelligent source, the mind of God.

Astronomer Edwin Hubble discovered the big bang in 1929. Physicists have found the light at the end of the preverbal tunnel. It's a different kind of light, the first light of God's creation. The big bang affirms the cosmos had a beginning, just as foretold in Genesis, God said, 'let there be light.' Genesis 1:3 The big bang is the preeminent smoking gun demanding the idea of a Creator be taken seriously. The discovery of the big bang reveals a whole new level of scientific ignorance.

At the request of Carl Sagan, Voyager 1 took its last photo of earth as it passed by Neptune in 1990. Sagan looked wistfully at the "pale blue dot." Like all of his contemporaries, Sagan believed the cosmos did not need a Creator because it had to be infinite. Hubble demonstrated a half-century earlier this assumption was wrong. If nothing else, this example clearly shows how hard it is to give up closely held beliefs, even in the face of indisputable evidence. Aristotle had proved by simple logic that the universe must be infinite because it is all that exists by definition. If the universe were bounded, we would want to know what is outside it. Without eyes, we would not know there is such a thing as light. In the same way, if we turn away from the awareness of God, it is as though He is not there. Objective reality is the eternal presence of God, which is experienced through spiritual awareness.

In 1925, Einstein discovered the law which proves the creation event was caused by nothing but information in the form of laws, *creatio ex nihilo*. Genesis tells us God is the Great Lawgiver. He spoke the laws by which we should live, and He spoke the laws that created and sustain the cosmos. Einstein also proved God's six-day forward-looking view of creation can be explained by relativistic time compression as the cosmos expanded at the speed of light. Today our backward-looking cosmological clock looks back in time 13.8 billion years to see the light from the big bang. In 1921, Niels Bohr proved atomic particles are a phenomenon created by waves of cosmic information.

What if God is not the cause? What is the cause? To say no cause is needed violates the first principle of science. Let us not forget that the rules of science only apply to physical entities. No one has ever claimed God is a physical entity. There will never be a scientific theory of how existence was created from non-existence. Existence and non-existence are not empirical, i.e., nor part of science, because they have no observable physical properties; science cannot even define something it cannot describe. Scientists who are atheists claim we are here due to an infinite chain of cosmic accidents. A scientific theory of accident is just an admission of ignorance.

No one knows how mindless atoms can create conscious beings. Why would mindless atoms want to get a life? How can it be that mindless atoms are studying themselves? No scientist can deny we live on a goldilocks planet in a Goldilocks cosmos. No one can deny that science has found a deep and meaningful relationship between the human mind and the cosmic order. We are more than a fleeting living form in a hostile cosmos. There is no mistaking we were meant to be here.

"If God created the cosmos, atheists derisively ask who created God?' Quite simply, finite minds cannot fathom the infinite. If there is no God, then existence has no meaning, and it would be correct to say we are nothing more than chemical scum. If life has no meaning, then words and deeds have no meaning. But words and deeds do have meaning. Since words and deeds have meaning, our lives have meaning; If our lives have meaning, then the universe has meaning. If the universe has meaning, it has a purpose, and we are here for a purpose. Being aware of purpose leads us to seek the origin of purpose, the mind of God.

Here is an example of how empirical knowledge and logic can impart an understanding of the necessity of God: Consider two hypotheses:

H1: Mindless accidents are the creator.

H2: An Infinite Mind is the Creator.

Argument:

- H1 is an empirical claim.
- Therefore, H1 must be a causal chain.
- By scientific principle, H1 cannot be its own cause.
- Every finite causal chain must have a cause.
- An infinite causal chain is empirically meaningless.
- Therefore, H1 cannot be an empirical claim. Thus, H1 is false
- H2 is not an empirical claim.
- H2 must be infinite to create all possible causal chains.
- Science has found the cosmos to be intelligible.
- The source of intelligible knowledge must be a mind.
- Therefore, H2 is a plausible hypothesis.
- Since H1 is false, H2 must be true.

The following argument connects humanity to the Creator:

- Only a mind can create laws.
- Laws govern the cosmos.
- The Mind that created the laws must be the Great Lawgiver.
- The Great Lawgiver of the Bible creates laws for a purpose.
- In the Bible, God declares humans are the purpose of creation.

In this discussion of evidence, God is on trial, you are the judge, and I am merely a public defender. The mystery of existence lies less in the observed material universe than in our capacity to comprehend it. Both science and religion have given us credible evidence of an unseen reality, which has more profound significance than material reality. If God is real, believers have nothing to fear from science. Science is useful, fascinating, and ultimately as mysterious as religion.

In 1988, physicist and atheist Leon Lederman received the Nobel laureate in physics. The title of his book, "The God Particle: If the Universe is the Answer, What is the Question?" [5] Like other physicists, he promotes the idea an infinite number of universes must exist in order to explain the existence of the one we know. Lederman does acknowledge the big bang is only one data point from which to draw his conclusion. The evidence concerning Lederman's question and his answer deserves a lot of attention because this is the speculative edge of cosmology. It does not invalidate the need for a creator. It's just more stuff needing a creator. The unknown is too vast to hold to the certainty of atheism. Einstein's words, published in 1930 by Science Week, "Glimpses of the Great," [6] provide a clear view of his beliefs. In response to a question about whether or not he believed in God, Einstein explained his view as follows: "Your question about God is the most difficult in the world. It is not a question I can answer simply with yes or no. But I am not an atheist. The problem involved is too vast for our limited minds. The human mind, no matter how highly trained, cannot grasp the cosmos. It seems to me, it is the attitude of the human mind, even the greatest and most cultured, toward God. We see a cosmos marvelously arranged, obeying certain laws, but we understand the laws only dimly. Our limited minds cannot grasp the mysterious force that sways the constellations." Saint Augustine expressed this sentiment 1,800 years ago. In a dream, he saw a little girl by the seashore, endlessly pouring pails of water into a little sandpit, never filling it up. This meant to him God will never be fathomed in this life. The Apostle Paul expressed the thought rather succinctly, "For now we see through a glass darkly." Corinthians 13:12 The paradox is how everything changes yet remains the same. The wisdom of the ages has grappled with the existence and meaning of a Creator. Now, more than ever, scientific knowledge weighs heavily on this Gordian knot.

I have learned from atheists to avoid God-of-the-gaps arguments. Science is capable of filling in its gaps. But science will never explain the infinite gap between God and us. There is no doubt science will eventually discover how life began. According to Genesis, the secret of life was included in the Cosmic Master Plan before creation even took place. Scientists cannot claim they have discovered the theory of everything until they can prove the theory predicts life. Both science and the Bible claim the cosmos is the result of a Master Plan. According to the Bible, God's only concern after creation was giving meaning to the life of those creatures He created in his image, *imago Dei*. Some will claim that if science ever figures out how life began, then the relationship between God and man suddenly becomes a God-of-the-gaps problem. This argument is specious because it removes the possibility that God could act within his own creation, making God impotent.

Because we have never seen God, the problem we face is like the ancient Indian parable of three blind men and the elephant. The blind men are asked to touch the elephant to learn what it is like. Each one gets to feel only one part, such as the trunk, tusk, leg, ear, or tail. They then compare notes and learn they are in complete disagreement. The conflict cannot be resolved, and bedlam ensues. In the real case before us, religious conflict tears at our ability to touch God. But God is never more than a touch away.

On the grand scale of things, no one can deny we are lucky to be here. The real question is whether we are here by dumb luck or are we lucky by design? Many people reject the "lucky by design" argument because too many bad things happen for there to be a benevolent God. The benevolence of God is manifested in the wonders of nature and the good works of those who love Him. The Bible does not promise us a rose garden. God told us we would have weeds in our garden and we should tend to it. If we are good and faithful gardeners, we may one day be chosen to walk with God in his perfect garden. "He will wipe every tear from their eyes. There will be no more death or mourning or crying or pain, for the old order of things has passed." Revelation 21:4

While science is incompetent to unravel the mysteries of the origin of the cosmos and life, it is exquisitely suited to explain their evolution. Evolution in nature is a goal-driven design strategy guided by physical laws to achieve the desired result, the glorious star-studded heavens, and our magnificent biosphere from nothing. It is a gross mistake to characterize evolution as a lucky accident. Make no mistake. Humans are extant evidence evolution was driven by some cosmic purpose. The design hidden in evolution has been revealed to the unlimited potential benefit of humanity.

According to America's greatest inventor, Thomas Edison, evolution is the method of genius. He said, "Genius is 1% inspiration and 99% perspiration." He tried over 10,000 filaments before he discovered tungsten works best in a light bulb. He said, "I have not failed. I have found 10,000 ways that won't work. Evolution does not "create" things. It is a way to make things from other things. As pointed out by Carl Sagan, "If you wish to make an apple pie from scratch, you must first make a universe." The principles of evolution are fundamental to the evolution of the universe from nothing, as well as capitalism, artificial intelligence, progress, stability, and ultimately, the very success of civilization. If human knowledge did not evolve, we would still be eating bananas in a tree.

Mystery of Existence

"Always be prepared to give an answer to everyone who asks you
to give a reason for the hope that you have." 1 Peter 3:15

Following the logic of Aristotle, scientists have always assumed the cosmos was infinite. The universe was infinite because it was believed to be all that existed, so a finite world made no sense. If existence were finite, we would want to know what exists outside existence. Scientists, like Carl Sagan, had concluded there was no need for a Creator. He wrote, "It is customary to say God created the universe out of nothing. We must ask the next question, where does God come from? If the answer is God always existed, why not save a step and declare the universe always existed?" The discovery of the big bang put an end to this argument. At the time, Albert Einstein wryly commented, "Only two things are infinite, the universe and human stupidity, and I'm not sure about the former." The discovery of the big bang revealed a whole new level of scientific ignorance. At the same time, the big bang is the preeminent smoking gun demanding the idea of a Creator be taken seriously. Thinking outside the cosmos is an entirely speculative endeavor. My contention is the trend of scientific discovery is far more favorable to the God Hypothesis than it is too atheism. Ignorance tends to cloud analytical thinking. The unknown is too vast to hold to the certainty of atheism. We now know the cosmos is finite, but existence, whatever it is, must still be infinite. But even the smallest possible fraction of infinity is still infinite. There is no place in infinity for our little cosmos. Our place in the grand scheme of things was expressed by Hamlet, "I could bound myself into a nutshell and call myself the king of infinite space." This logical non-sequitur is an intractable dilemma that can never be solved by science. But it leaves plenty of room for an infinite or eternal God. Our minds are the only room He cares to occupy n our pathetic little world.

To this day, inquiring minds still want to know precisely what does it mean to "believe in God?" The answer is simple to explain, but at the same time, God is a very subtle mystery. God said to Moses, "This is what you are to say to the Israelites, YHWH, (pronounced Yahweh, meaning, "I am that I am"). This is the name by which I am to be remembered from generation to generation." Exodus 3:14 God identifies himself as the eternal Infinite Being who has always existed. Physicists now know it is physically impossible for anything physical to have always existed. God is our only connection to the infinity of existence.

Existence has no physical properties. It just is. It is the ethereal nature of reality that makes us believe something is real, rather than not real. We are real only because we think we are real. God becomes real only when we believe He is real. What we find is real is just in our minds. If existence is real, we should want to know what does existence mean outside our minds. If we think what is there is real, it must exist in God's mind because all that exists is contingent on the great thought that always existed. Our minds are connected to God's mind, by the handiwork of his creation and our existence as an eternal being made in the image of God. C. S. Lewis wrote, "You don't have a soul. You are a soul. You have a body." [3]

Science cannot tell us anything about existence, the infinite, the state of being, consciousness, or the soul. But Physicists have known for over a century that material reality is only an illusion created in mind. [Physics professor Michio Kaku stated, "Quantum physics is not what you think it is. Maybe you are real, but everything else is fake. In the matrix world, reality can be reprogramed."] Physicist Carlo Rovelli wrote the book, "Reality Is Not What It Seems." [7] Modern physics strongly suggests we exist within the reality of God's mind. Only God has the power to create or destroy our eternal souls. All other things are figments of our collective imaginations, like a mirage we share. Logic is limited to the finite. Logic can only get you from point A to point B. Imagination can visit any path lost in your personal universe in order to steer you to your destiny. Imagination allows you to see beyond the finite to the infinite universe, the mind of God.

The secrets of reality transcend the illusion of the material world, to the ethereal quantum world, to the unknown world outside our created world, to the symbolic worlds inside our minds, to the infinite world of all the worlds, the eternal world of the mind of God. Science has opened a new window into the ultimate mystery we call God. Reality is not reality as it really is, or indeed ought to be, or really could be. The deep and meaningful relationship between the human mind and the cosmic order can hardly be an accident. The mystery of existence lies less in the observed material universe than in our ability to understand it.

Both science and religion have given us credible evidence of an unseen reality, which has more profound significance than material reality. The first big hint is how small and insignificant we are compared to our genuinely unique importance as conscious, intelligent beings. This is the core of the human condition. We have a mind made in the image of God while trapped in a dirtbag. Science has opened the door to a new vision of reality that defies our collective common-sense experience.

Cosmic Miracles

"Either nothing is a miracle or everything is a miracle." Einstein lore

"If you believe in miracles, all is explained. If you don't believe in miracles, you better hope you never need one." Yours Truly

Humans can now see what outer space looks like. Thank you, NASA. You have no clue how smartphones work. Thank you, engineers. You don't know how electronics work. Thank you, physicists. When you read about physics, don't panic. You don't need to be a rocket scientist. All you need is a good imagination to appreciate the fantastic discoveries of physics. They are a guide to our destiny. Here's a look at where we are headed.

Humans started learning about physics when they picked up rocks and watched them fly. Today, we all have an intuitive understanding of fundamental physics. We know why we don't jump off buildings. We instinctively know Newton's laws of motion. It takes a force to put an object in motion. An object in motion tends to stay in motion. Energy is transferred when objects collide. Don't fear. Children know why helium balloons take flight and water runs downhill. It's just physics.

Isaac Newton and Albert Einstein are recognized as the greatest physicists who ever lived. The inspiration shared by them transcends science. Both viewed science as a sacred quest to know the mind of God. To them, the discovery of the laws of physics was a miracle; it can only occur once and cannot be repeated. In his own inimitable way, Einstein explains, "One may say, 'the eternal mystery of the world is its incomprehensibility. It is one of the great realizations of Immanuel Kant that the postulation of a real external world would be senseless without this comprehensibility. In speaking here of 'comprehensibility', the expression is used in its most modest sense. It implies the production of some sort of order among sense impressions, this order being produced by the creation of general concepts, relations between these concepts and by definite relations of some kind between the concepts, and sense experience. It is in this sense that the world of our sense experiences is comprehensible. The fact that it is comprehensible is a miracle." [8] Einstein envisioned a theory that would explain everything. He was never able to put it to paper, but physicists today are inspired by his vision.

Physics presents us with a startling future. Unfortunately, you will not be here for the 2134 Olympics. However, the generation being born today will walk the streets with machines, play a role in their local 4D holographic simulator, vacation on the moon, live on Mars, explore exoplanets, or even communicate directly with their minds. The human race is on the verge of migration into the final frontier, the cosmos.

Meanwhile, back on earth, science is busily following the yellow brick road to our destiny. This is not an unguided walk in the dark. The path of science is sparkling with majestic beauty better than diamonds and gold. The laws of physics are an elegant and magnificent triumph of human intellect.

The laws of physics have guided the evolution of the cosmos for 13.8

billion years. Discovering these laws is a testimony to our vaulted place in the universe. Knowing the laws of nature is the key to the future and vital to the very survival of the human race. As you find out about these laws, keep in mind laws can only be created by an intelligent mind and their magnificence is just a hint of the glory of the Mind who created them. And how can we know this? We shall find the laws of the cosmos must have existed before time began. If this were not the case, our finite universe could not be here. Rules do not create themselves. Laws require an intelligent Mind by definition. To know in your own mind that God is the ultimate Intelligent Mind is the miracle of all miracles! Those who cling to the mindless materialism of the Victorian age are missing out on the most significant revelation of God in more than 2,000 years. Most people marvel at the stars, but few remember to thank God for the excellent view. "Miracles are a retelling in small letters of the very same story, which is written across the whole world in letters too large for some of us to see." [3] C. S. Lewis, British author. We should be thankful for a wonderful world that provides for all our needs. Que mundo marvaloso!

Everything that is too good to be true is a miracle. A mother with child is the miracle of life. We all share in the miracle of life, beginning from a single cell. Let me be very clear. Science is our most valuable tool when it comes to understanding our world. But a scientific understanding does not necessarily negate miracles. First of all, an explanation by itself does not negate a miracle. Healing lepers was a common miracle by Jesus which has been eradicated by modern medicine, so no one knows what the future may hold. Even more relevant, the scientific method requires an event or sequence of events to be repeatable. If this is not the case, science fails to be a valid explanation. The beginning of life is a miracle because it only happened once and has never been repeated in nature nor by science. Since every single living thing, whether it be a plant, animal or human, is the result of a very specific and unique sequence of genetic changes, meaning every life is a miracle. In fact, our entire universe is a miracle because it cannot be repeated. The creation of the cosmos and the origin of life are miracles. An ordinary day for us would seem to be full of wonders to people who lived before the modern era. The greatest miracle is the human mind can fathom the mystery of the Infinite Mind of the Creator.

Common sense and rational thinking lead us to be skeptical of miracles. And science has put a fine point on what is and isn't a miracle. This is where faith begins. Doubt will always be a part of faith. But as you contemplate the wonder of it all, you should start to realize everything about our existence begins and ends with faith tempered by reason.

If Jesus is who He said He is, all his miracles are explained. Or did the story writers take too much liberty with the truth? Was Jesus a master hypnotist or charlatan? You are entitled to your own beliefs, but you should be able to explain it to others in a way that makes sense.

Over the course of his three-year ministry, Jesus averaged a miracle every month or two, each one involving different laws of nature and different people., including large crowds. The astonishing number and the diversity of miracles recorded is anecdotal and circumstantial evidence to consider. But these miracles require substantiating reasons and facts. Most of the miracles attributed to Jesus are common place today due to the miracle of modern medicine. Even evolution is a miracle because each step is not a miracle, but the path itself is a miracle because it can't be repeated. The laws of classical physics will not change, but quantum physics suggests the power of the human mind can make local exceptions.

"I know of nothing but miracles." Walt Whitman, American philosopher.

CHAPTER 2

ANCIENT WORLDVIEW

Primeval Torah

"In the beginning, God created the heavens and the earth." Genesis 1:1

The first 11 chapters of Genesis are known as the primeval Torah. These passages represent oral tradition, before the written word, of the Jewish people from 4,000 BCE to the time of Abraham, circa 2,000 BCE. The legend of the great flood at the time of Noah is also recorded in Babylonian history, only the Babylonian story features a giant woven basket, rather than an ark. Around 7,000 BCE, the Mediterranean Sea breached the Bosporus Straight. Ancient villages were left submerged under 300 feet of water, giving rise to speculation that this event is the basis for the Noah's Ark story. Legends like this represent the beginning of monotheism, the greatest religions known to humankind. While the great civilizations around the region had formidable armies and worshiped the sun, Israelites knew the sun was just a big light; they trusted God as their protector. Religious customs consumed their culture. If the spiritual truth is to be found, their Bible is the first place to look. Religious truth confronts the more profound questions of ethics and the purpose of life. The godliness of the Jewish people is expressed as a well-crafted morality lesson in every passage of their Bible.

If God is the titanic answer to our existence, the scientific credibility of the entire Bible rests on the small shoulders of the first chapter of Genesis. The six day creation story is a simple and elegant account of how we got here. At first glance, the story is notable because creation is not some one-time event as would be expected from ancient folklore. Instead, it is a well thought out sequence of events over time, much like modern reasoning has found. But it continues to fuel a culture clash between science and Christianity, which won't go away.

In the seventeenth century, Bishop Ussher used the genealogy table in Genesis to determine creation occurred at 9 a.m., October 27, 4,004 BCE.

According to God's creation clock, the beginning of time to the final end of time occurs in seven days. A day is the shortest time span of the cosmic clock, as understood by the Israelites. During this one week of time, God is watching the universe evolve from nothing at the speed of light before him. Time for God is at a near stand-still. This phenomenon is explained by Einstein's theory of special relativity. Time is not what you think. Time is different for everyone and every place. For example, a watch on the second story will run faster than a watch on the first floor, gaining one second per year.

Gerald Schroeder, an eminent physicist, and Hebrew scholar, deserves great credit for solving the mystery of the six days of creation, resulting in humankind. During this time period, astrophysicists have observed the stretching of space and time billions of times, especially during the earliest period. The stretching of time has slowed the passage of time for observers within the universe to 13.8 billion years. The universe can be visualized as contained on the surface of a stretching sphere, such as a balloon. As the balloon stretches, all the galaxies on its surface are moving away from each other. The left-over radiation from the big bang envelopes the balloon's surface, even today. Because light travels at a finite speed, astronomers cannot observe other galaxies and other celestial objects until they can be seen within our event horizon, as illustrated below.

The event horizon is determined by the time it takes for light to reach the earth from somewhere in the cosmos. As shown in the diagram, the event horizon opens up with time and distance to give cosmologists a complete view back to the creation event. When we see a light beam emitted by a star, we are seeing exactly how that star looked when the light left the star, the exact same view as God sees instantly as it happens. The information from the star is perfectly preserved, even though the light has been traveling for billions of years to get here. Time within the light beam has stood still. The light from the big bang provides an instant record of the future of cosmic evolution. Nothing can escape from God's knowledge, not even the number of hairs on your head. Because we are living in the sixth millennia of the seventh and last day of God's clock, it can be surmised from the table below that time for us will expire; It's only a matter of when. Will mankind be given time to explore and colonize the universe; only God knows.

The credibility of the Bible is under the microscope of modern science. There is no wiggle room in the search for truth. Dr. Schroeder has put an end to the creationist's "young earth" theory.

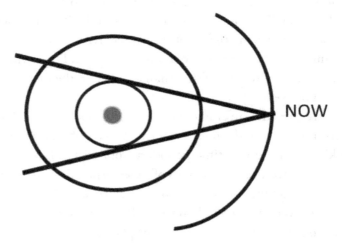

Expanding Universe intersected by Event Horizon

After God created his soul mates, Adam and Eve, passing time in the Bible is denoted by a yearly calendar of human life spans. In Deuteronomy 32:6, 7, Moses says, "Is He, not your Father, your Creator, who made you and formed you? Remember the days of old; consider the long generations past. Ask your father, and he will tell you." Time is clearly distinguished into two periods, days of old, the period of creation, and generations long past, the time elapsed since the creation of mankind, measured in years. Another insight into the nature of the creation clock is given in Psalms 90:4, "For

1,000 years in your sight are like a day just gone by, or like a watch in the night." 1,000 is the largest number in use by the Israelites, so its use here means unaccountably large. This verse foretells the great mystery of cosmic time discovered by Einstein. Schroeder used scientific analysis to describe how God's six-day creation clock is entirely consistent with our cosmic clock, which looks back billions of years to the big bang. As told in Scripture, the creation story is not history; it is written as if the events are unfolding in real time, the God's eye view of events is time compressed by special relativity. The laws of physics indicate time is not a fundamental property of the universe, suggesting God's master plans for the universe were initially intended to be eternal, without time. This fatal demise of God's original plan for mankind is first chronicled in the second chapter of Genesis, followed by God's displeasure with humanity right up to the end of time in Revelation.

While God was revealed through Scripture, thousands of years ago, the secrets of creation have been hidden in nature until now. Through science, God has given humanity the evidence and reasoning power to understand the mysteries of his creation. The most recent scientific theory of creation has been called the "Creator's Voice." Expect to be amazed by the revolution taking place in modern physics. We are very fortunate to have the history of the cosmos revealed to us. Is this just a coincidence, or has our generation been chosen to see the works of God as King David could only dream about?

The Genesis story has been criticized because water appears before the creation of the earth. In actuality, water in the form of comets was produced by UV radiation during the formation of our galaxy. During the formation of the earth, hundreds of thousands of comets rained down on earth. This bombardment lasted billions of years, and the resulting water world and cloud cover would have persisted at least that long. This is entirely consistent with creation day two of the creation timetable below. Also, although the sun and stars existed before the earth they were not visible from the earth, where God's Spirit dwelled. "The Spirit of God was hovering over the waters." Genesis 1:2 The Spirit of God literally planted the seed of life in the waters. On the third creation day, the persistence of dense creation cloud cover, fog, and volcanic ash lifted, clearing the sky. This condition is consistent with God's focus on the creation of his unique planet.

1

The mention of morning and evening before the creation of the earth is another criticism leveled by skeptics. Again, the relevant information is lost in translation. Schroeder explains each day of creation begins with evening (the Hebrew word, eres), and ends with morning, (the Hebrew word, boca). This is often considered erroneous by critics because the sun is not even mentioned until the third day. But these are derived Hebrew words whose root meanings are much more profound. The root word for eres means chaos or darkness; the root word for boca means order or light. This description of time is unique to the creation story, and the meaning is clear. Over an extended period, God brought forth order from chaos as darkness moved into light.

It is also important to note the Bible mentions prehistoric time occurred on day four. Genesis 1:20, 21, "God said, 'Let the water team with living creatures and let birds fly above the earth across the expanse of the sky. So, God created the great creatures of the sea and every living and moving thing with which the water teams, according to their kinds, and every winged bird according to its kind." Early humans would not have known of dinosaurs because most lands like the Middle East were prehistoric sea beds, so there would be no fossil record of land creatures to be found. But skeletons of extinct sea creatures may have been found, which were not fish. But somehow, they knew birds existed in the prehistoric period before common land animals appeared on day five.

The Ichthyosaurs (sea lizard) was the top marine predator all over the world for 200 million years. These air-breathing sea monsters had teeth like knife blades, eyes the size of a human head, and grew up to 50 feet long. They survived for 300 million years. They became extinct 90 million years ago due to toxic undersea volcanic activity. Skeletal remains of these creatures may have led to the legends of dragons throughout early history. In contrast, the top land predator, T-Rex, lived on an isolated landmass, now North America, for only two million years.

Creationists reject evolution because the story says God created creatures "according to their kind." In actuality, evolution has mechanisms that prevent cross-breeding.

There are scriptures in Genesis that make it clear we are living in the period of creation day seven; God is currently resting. The following table connects the event sequence in Genesis to corresponding times of events as discovered by science.

Time zero	13.8 billion years ago	light/matter
Day 1	8 billion years ago	galaxies/ comets
Day 2	4.5 billion years ago	earth/oceans/plants
Day 3	550 million years ago	sky clears/Cambrian explosion
Day 4	150 million years ago	great sea creatures/birds
Day 5	50 million years ago	mammals/animals
Day 6	50,000 years ago	homo sapiens
Day 7	current time	God rests

The story of creation in Genesis is like the picture on the lid of a puzzle to be solved by science. Everything we have learned from science fits together like the pieces of that puzzle.

History Begins

"Then God said, let us make man in our image." Genesis 1:26; The use of a possessive plural term for God is not used again nor clarified in the Hebrew Bible.

Our unique attributes that differentiate us from chimpanzees evolved for seven million years. Further back, we represent modifications of great ape attributes that are 10 million years old, primate attributes are 50 million years old, reptilian characteristics are 250 million years old, vertebrate characteristics are 500 million years old, and characteristics of nucleated cells are 3.5 billion years old.

10 million years ago, the earth's climate entered a colder and drier phase, which brought on the ice ages beginning 2.5 million years ago. The North African tropical forest began to recede, being replaced first by open grasslands and eventually by desert, the modern Sahara. Some primates adapted to ground dwelling.

Over 26 distinctly different skeletal remains trace the evolutionary development of the humanoid species. In 2010, fossilized animal bones bearing marks from stone tools were found in the Lower Awash Valley in Ethiopia. Discovered by an international team led by Shannon McPherron, at 3.4 million years old, they are the earliest evidence of stone tool use found anywhere in the world. Hominin living in Ethiopia 300,000 years ago discovered how to shape volcanic glass into sharp points for spears and knives. The most unlikely species became the most lethal hunter at the top of the food chain. Like us, they were slow, weak, and devoid of sharp teeth and claws. The transition from herbivore to carnivore brought a high protein diet that hastened their evolutionary climb out of nature, powered by an unusually large brain that is three times the size of our nearest relative.

At the peak of the last Ice Age, 70,000 BCE, climate change started the great migration of 10,000 primitive people. By 25,000 BCE, the Neanderthals became extinct. Our ancestors, the Cro Magnons, had migrated around the globe.

Human intelligence did not evolve primarily as a means of survival in their environment, but rather as a means of surviving in large and complex social groups. Behaviors for living in large groups include cooperation, generosity, competition, deception, and coalitions. These group dynamics relate to the ability to understand the thoughts and emotions of others. When the size of a social group increases, the number of different relationships in the group typically increase by orders of magnitude. Chimpanzees live in groups of about 50, whereas humans usually have a social group of about 150 people, which is now referred to as the Dunbar number. "According to the social brain hypothesis, when humans started living in large groups, selection favored greater intelligence. As evidence, British anthropologist Robert Dunbar has demonstrated a relationship between neocortex size and group size of various mammals." [10]

Ancient humans were a superstitious bunch, just like today. You will never find a thirteenth floor in a hotel. With great imagination, prehistoric humans across the globe crafted over 1,000 mythical creation stories. The reasons that compelled them to do this are with us today; it has been passed down in our genes.

Humans saw the need to invent gods to explain what they observed in nature. As early humans learned about life and nature, a rudimentary form of religion and science came together in every region of the world. The forces of nature were personified as gods with great powers. The sun made day and night. The sun's movements told the seasons. The moon, stars, and planets formed a calendar. Eclipses of the sun and moon, and comets and shooting stars, were terrible omens. The heavens were the overseers of the world. Religious artifacts and astronomical stone markers were commonplace. The worldwide human population reached six million at the beginning of the Bronze Age, around 4,000 BCE. Over the next millennium, the foundations of Western Civilization emerged in the Near East. The Egyptians and Babylonians harnessed the waters of the Nile, Tigris, and Euphrates Rivers to support agrarian societies. They became great empires, building massive temples, pyramids, and monuments. They developed wheeled transportation, warfare, writing, literature, law, mathematics, astronomy, and the arts. In the barren lands between these two empires, a small tribe of nomadic sheepherders, known as Israelites, would become to be known as God's chosen people. The epic biblical story of Moses dates to the thirteenth century BCE. The book of Exodus records the struggle between the all-powerful God of Israel and the pharaoh of Egypt, thought to be Ramses II.

In the eighth century BCE, Solomon's Temple was built in Jerusalem. The innermost sanctum, the "Holy of Holies," held the Ark of the Covenant with the Ten Commandments on stone tablets. The first five books of the Bible, the Torah, were written by priests near the end of Solomon's reign, although tradition attributes the writings to Moses.

During this period, a Greek poet, Homer, wrote the "Iliad and Odyssey." The bigger-than-life tales chronicles the mythical exploits of heroic men who would come to symbolize the ideal man in Western culture, a beautiful mind in a beautiful body. But the anthropomorphic Greek gods were not up to the task of competing with the God of Israel. The absurd little Greek gods had human foibles and capriciously meddled in human affairs. The culture clash between Greek humanism and Jewish monotheism set the path of Western Civilization.

The Hebrew Bible recounts the history and culture of the Israelites up to the fifth century BCE, not long after Solomon's First Temple was destroyed by the Babylonians. The fifth century BCE also marks the high point of classical Greek culture, known as the Golden Age of Greece. They gave us great architectural monuments, literature, poetry, theater, music, art, higher learning, democracy, competitive sports, philosophy, and science. These are the building blocks of Western Civilization.

The father of philosophy, Socrates, is best known for the Socratic method of learning by inquiry. He developed the technique of hypothesis testing by a process of elimination by contradiction. He said, "The unexamined life is not worth living." He was sentenced to death for belittling his peers. His protégé, Plato wrote, "Socrates pondered the possibility of the afterlife before drinking poison Hemlock."

Plato defined knowledge as justified true belief. He allowed convictions to count as knowledge even if they could not be proved. A priori (prior to) reasoning, mathematics, and geometry could only be known in his metaphysical reality of perfect expertise. Plato's philosophy of metaphysics clearly shows his familiarity with Jewish ideas of a spiritual reality. He imagined the material world was just the shadowy image of a higher absolute truth. He reasoned a chair could take many forms in the material world, but a universal mathematical or geometric abstraction can represent all chairs. Since mathematics and geometry can describe nature in the abstract, Plato concluded the power of mathematics and geometry transcended nature. Therefore, there must exist a higher reality of universal ideal "Forms" that can be known through the mind. The mental truth of Plato had more profound meaning to him than mere mathematics. Plato envisioned a reality of such objective sense it transcends all human thought. The existence of the Platonic world cannot be deduced into a reality by logic alone. But it does provide a simple picture of a comprehensible cosmos consistent with all that can be known from observation and logical analysis. The existence of the Platonic world of metaphysics would have to wait until physicists discovered Plato's vision provided a framework for a mathematical theory of everything!

Plato's protégé, Aristotle, developed the system of formal logic in use today. He studied plants and animals to organize them into a classification system that would become the starting point for biology. He reasoned the rate of fall of objects was proportional to their weight. 2,000 years later, Galileo proved him wrong by dropping different size cannonballs off the Tower of Pisa.

Pythagoras developed rules of algebra and geometry, and invented formal proofs. He also discovered the importance of mathematics in music. For example, two notes an octave apart differ by a factor of two in frequency.

Science was recognized as a way to bring order to a chaotic world and harness the forces of nature. The Greeks believed all that exists can be classified as earth, fire, air, or water.

Philosopher Democritus began his famous discoveries when he was sitting around wondering how many times he could cut a loaf of bread in half. He reasoned, no matter how many times it is cut in half; there must always be something left. The final piece that could not be cut in half, he called the *atomos*, which means indivisible.

Philosopher Zeno is best known for the "Achilles and tortoise paradox" in which he uses Democritus' method to prove that motion is just an illusion. In a thought problem, Achilles gives the tortoise a commanding lead in a foot race. In each increment of time, each contestant moves halfway to the goal line, the paradox arises when Achilles cannot move past the tortoise. The only way Achilles can win is to make a "quantum leap" to the finish line. Therefore, motion is both continuous and quantized, hence the paradox.

With a solid foundation in mathematics, geometry, and logic, natural science came into its own in Greece. They learned a great deal from astrological observations of the Babylonians and Egyptians. Astronomy has led the way to the understanding of the cosmos. What the ancient Greeks learned without a telescope is a monumental achievement of discovery. They figured out the diameter of the earth, moon, and sun. They determined the distance to the moon and sun. They hotly debated whether the earth orbited the sun or the sun orbited the earth. As we know, two eyes use parallax to determine distance. In the first century CE, Ptolemy decided the position of the earth was fixed because he observed no parallax when looking at stars with his naked eyes.

Eratosthenes was the head of the vast library of Alexandria. He used the relative angle of sun shadows at two distant locations to measure the curvature or circumference of the earth to an astounding one percent accuracy. He also measured the tilt of the earth relative to the sun, and he invented the leap year to improve calendar accuracy.

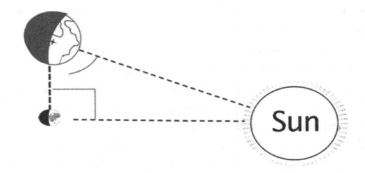

The diagram illustrates how Aristarchus, knowing the distance to the moon, calculated the distance to the sun using trigonometry. The angle from the earth to the sun at new moon and the right angle at half-moon form the right triangle. He used the relative size of the moon to earth during a lunar eclipse to form similar triangles that could then be used to measure the size of the sun.

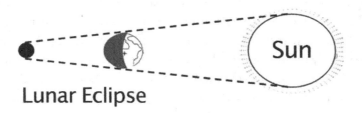

Lunar Eclipse

The 30-year Peloponnesian wars between Athens and Sparta brought the Golden Age of Greece to an end in 400 BCE. The Ancient Library of Alexandria was the largest and most significant in the ancient world. It stood for three centuries until the Romans conquered Egypt and burned it to the ground in 30 BCE. The Pope ended attempts to recover some of the loss in 341 CE. Saint Augustine recognized the value of Plato's work and tried to bring it under the Christian umbrella through the Franciscan monks. But the conservative Benedictine Order ultimately forced purification of Church doctrine. The catastrophic loss of knowledge from the great ancient empires stands as a symbol of cultural annihilation.

Rome established a republic with an emperor and a senate in 500 BCE. Nine years before Jesus was born, Julius Caesar became the emperor. The month of July is named after him. During his reign, the Roman empire grew to include most of Europe. Seeing the need for greater leadership power, he declared himself emperor for life, converting the Roman Republic into an empire run by a dictatorship. Jesus brought his ministry during this pivotal time in human history. Two years after Jesus was crucified by the Romans, Julius Caesar was assassinated by a close advisor, saying *"et to Brutus"* as he lies dying. During the following century, more than a dozen emperors would be assassinated. (Speaking of tipping points, we are due for the Antichrist to bring peace to a troubled world.)

Common Era

The birth of Jesus marks the beginning of the Common Era. Jesus grew up in a small Jewish village. It was not kosher that he was the bastard child of a very young, supposedly virgin mother. All we know is He was a carpenter like his stepfather until He was 30. He spent 38 months as an itinerant teacher by the Sea of Galilee, a backwater of civilization. His ministry served the low class, sick, poor, women, and slaves. He was very critical of the rich and powerful. His brothers did not believe his teaching. His 12 disciples were mostly illiterate fishermen. Jesus was often quoted, but He himself never wrote a word. His disciples could not grasp the most absurd claim anyone has ever made; He was the Spirit of God in a man's body. He was tortured and crucified for blasphemy. His followers denied knowing him out of fear. If the story ended there, it would have been relegated to the dust bin of history. But then something happened. His closest followers swore to their death He arose from the dead. This was a most astounding claim because there is no record of the Jewish religion ever making such an outlandish claim. The Jews rejected Jesus as a false Messiah. People scoff at his story, but it is the foundational belief of Christianity. We can only conclude Jesus was who He said He was, or his followers pulled off the greatest hoax in history.

In 67 CE, Shimon bar Kosha was declared the true Messiah when he led a successful revolt against the Romans, establishing the independent Jewish state of Israel. When his state was conquered by the Romans three years later, every stone of Solomon's Temple lay scattered, and the rabbis called their Messiah, Simon bar Korea, Son of Lies.

A rabbi named Saul had an epiphany as he traveled to Damascus to persecute followers of Jesus. Though he never met Jesus except in a vision, this highly educated rabbi spread the gospel throughout the Roman Empire, under the Greek name Paul. Ten of Jesus' disciples and Apostle Paul willingly died martyrs for their beliefs. For three centuries, untold tens of thousands of Jesus' followers fearlessly braved torture and death rather than swear allegiance to the emperor as the greatest of all gods. The good works of the early Christians among the sick and poor would not go unnoticed.

At the beginning of the third century, Emperor Constantine's mother converted to Christianity, which soon became the official religion of the Roman Empire. Near the end of the fifth century, the Romans fell at the hands of barbarians.

For 1,000 years, Europe fell into the feudalism of the Medieval Age. The absence of governments gave the Roman Church supreme power and authority over Europe. Monasteries were the only institutions of learning. The Church became very wealthy and erected cathedrals in the center of every town and village in Europe. Monasteries became the only institutions of education.

Medieval Apologetics

In the twelfth century, Anselm, the Archbishop of Canterbury, founded a field of scholasticism called Christian apologetics. It is the philosophical defense of religious beliefs using reasoned arguments. Anselm's ontological argument asserts, "God, by definition, is for which no greater can be conceived. God exists in the understanding. If God exists in the understanding, we could imagine him to be greater by existing in reality. Therefore, God must exist." His best argument reads as follows; "If something can be conceived not to exist, then something greater can be conceived. Consequently, a thing, which nothing greater can be conceived cannot be conceived not to exist and so it must exist." This argument has content when we consider a finite cosmos, which can be conceived not to have always existed.

A more elaborate ontological argument was developed by Gottfried Leibniz, a German mathematician, and philosopher of the seventeenth century.

Apologetics concerns itself primarily with the existence of God according to the following argument categories:

- Ontological -the concept of God implies God exists.
- Cosmological -the existence of the cosmos suggests a Creator.
- Teleological -a purposeful cosmic design requires a designer.
- Moral -objective moral values imply an absolute Source.
- Transcendental -our mind implies a greater Mind exists.
- Presuppositional -fundamental beliefs concerning human nature suggest God as a prerequisite.

Thomas Aquinas was a thirteenth-century Dominican friar who attempted to synthesize Aristotelian philosophy with the principles of Christianity. He defines four kinds of law: eternal, natural, human, and divine. Eternal law is the decree of God that governs all creation. It is, "that Law, which is the Supreme Reason that cannot be understood to be otherwise than unchangeable and eternal." Natural law is human participation in the eternal law and is discovered by reason. Natural law is based on the "first principle" that good is to be done and promoted, and evil is to be avoided. All other precepts of natural law are based on this. Natural and human law are not adequate because human behavior needs direction from the Divine law revealed in Scripture. The writings of Aquinas are core Catholic curriculum for the priesthood, and he is considered to be the Church's greatest theologian and philosopher.

Medieval apologetics predates the advent of science and relies on arcane logic, which has not held up to the test of time. As will be shown, science is literally the final say in modern apologetics because the latest scientific discoveries now completely confirm for all time the Truth of the Bible in exquisitely glorious detail.

Renaissance

The power and wealth of the Church eventually led to intolerable corruption. In 1492, Pope Alexander VI of the Borgia dynasty would become known as the most corrupt Pope in history. ("The Family" by Mario Puro is a novel about him.) That same year, Columbus discovered America.

He knew the earth was round, but his calculation of the circumference was five times too small.

Italy benefited from its position on the trade route from Europe to the East. The Medici Bank of Florence became the largest in Europe and the Medici House was the wealthiest family in Italy. They became concerned that Jewish law prohibits usury. In order to get through the Pearly Gates, so the Medici's lavished their wealth on the Vatican. Their wealth underwrote the High Renaissance in Italy. Renaissance, French for "rebirth" is the period in European history from the fifteenth to the sixteenth century. It would lead Western Civilization into the modern era. The fifteenth-century brought a restoration of the great Greco-Roman culture of the ancient classical period.

Two geniuses, Leonardo da Vinci and Michelangelo, would come to epitomize the archetypal humanist ideal, the Renaissance Man. Da Vinci was a prolific artist and inventor. His Mona Lisa is the most famous painting in the world. He carried it with him for 30 years until his death in Paris. Michelangelo was a sculptor, artist, and architect. His style is both erotic and compassionate on a grand scale. His statue of David is arguably the most important cultural icon in the world.

Like many of their peers, these men studied ancient Greek culture. They knew they could not achieve artistic greatness without detailed knowledge of human anatomy. They had to perform their anatomical work in secret because Church law forbad desecration of a dead body. They had to conceal their admiration for Greek art from the Vatican under the threat of ex-communication or even death. The statue of David hides his identity as the ideal Greek man. David knows God is on his side, yet he has confidence in his skill. David is the beautiful man in a beautiful body doing the work of God. He is an iconic symbol of the Renaissance Man. He is the model for all humanity. His struggles to please his God are well known. His humanness failed before God, but God loved him more than any other man because God looked upon his soul.

God said, "I have found David a man after my own heart." Acts 13:22

The 14-foot statue of David by Michelangelo was commissioned in 1500 by Pope Alexander VI to be the pinnacle of the Vatican Basilica. It never made it up there. Instead, it stands in pristine condition as the centerpiece of the Academy Museum in Florence, Italy. The greatness of the Renaissance man as a symbol of Western Civilization is ingrained in a single marble stone. David is the ideal man of the ages, ready to do the will of God.

History is more than just "one damned thing after another." Historian Arnold Toynbee said each civilization was a response to a challenge. In the sixteenth century, the Italian Renaissance revived ancient culture and knowledge to replace a spiritually bankrupt Roman Catholic Church. Classical Greek architecture, humanities, arts, music, and literature were restored. Universities flourished as intellectualism blossomed. The power of the Roman Church would give way to the new world order. The Western world began the transition into the modern era. As nation building emerged in Europe, the dominance of the Catholic Church was threatened. In 1440, the invention of the Gutenberg press would soon put a Bible in the hands of the common people. In Germany in 1517, Martin

Luther, a Catholic priest, posted 95 theses against the Church, which began the Protestant Reformation. The invention of the Guttenberg press and the translation of the Bible into common languages brought religion to the common folks. The Protestant Reformation swept through Northern Europe. John Calvin published an "Instruction Guide" for Christians and started the Baptist Church. John Knox brought Calvinism to Scotland. In 1534. Henry VIII established the Church of England because the Pope wouldn't give him a divorce. John Wesley began the Methodist Church.

Science emerged in Western Civilization because of two important influences. First, Greek philosophy recognized man could come to an understanding of his world through geometry, mathematics, and reason, and second, Judaism brought us the idea of the Supreme Law Giver who created the universe by his spoken word and structured it in a way would fulfills his purpose. For centuries, natural philosophers were convinced there was order in nature, but it was hidden from humans. It can't be seen in our daily lives. The notion human beings could understand nature, literally read the Mind of God through reason and inquiry, is the spark that motivated the birth of science. During the Middle Ages, all Western knowledge was held within the monasteries of the Catholic Church. The great thinkers of the time were theologians who started ecclesiastical institutions, which led to the system of universities of higher learning we have today. They began to transform the natural philosophy of the ancient word into modern science during the sixteenth century. An orderly investigation of nature was a religious matter of great importance.

The year was 1543 when Copernicus, a distinguished Polish scholar, found a way to study the cosmos, which had been lost for nearly 2,000 years. It would be the spark that ignited the scientific revolution. Science would soon take its place at the foot of knowledge. Copernicus said, "Science is more divine than human."

European culture progressed into the Age of Enlightenment in the seventeenth century. The Industrial Revolution brought wealth and other social benefits to the working class. Worn out traditions were replaced by humanism, reason, critical analysis, and individualism. A tidal wave of change swept social and religious upheaval across Europe, including Protestant-Catholic wars, the American and French revolutions, and eventually World War I, and the Russian revolution.

CHAPTER 3

AGE OF ENLIGHTENMENT

Copernican Revolution

"The star keeps all the planets on their appointed orbits, yet somehow manages to ripen a bunch of grapes as though it had nothing else to do." Galileo.

In the second century CE, Egyptian astronomer Claudius Ptolemy constructed the first mathematical description of the solar system. It required very complex geometry to place the earth, rather than the sun, at the center of the planetary system. As shown below, this representation of the heavens was universally accepted for 15 centuries.

Fifteenth century universe

Nicolaus Copernicus (1473–1543) was a Polish physician, attorney, governor, mathematician, and astronomer of the Renaissance period. He formulated the first scientifically correct mathematical representation of the solar system, the heliocentric model. Fearing religious persecution, he delayed publication until near his death in 1543. In the Copernican system, the moon was not a planet; it was a satellite of the earth! The Copernican system illustrates an important scientific principle called Occam's razor. William of Occam was an English friar who reasoned the simplest explanation of natural phenomena is the best unless it can be proved otherwise.

It wasn't long before the work of Copernicus was placed on the index of banned books in Italy. The Roman Inquisition would eradicate godless ideas. A century after Copernicus died, the consummate Italian scientist, Galileo Galilei, became a champion of his nearly forgotten work. Citing this work, Galileo said, "The universe is written in the language of mathematics." Galileo built one of the first telescopes and observed the magnitude of stars, the craters on the moon, four of Jupiter's moons, the rings of Saturn, and sunspots. Because the heavens were the perfect sphere of God, the Vatican declared it was heresy to look through a telescope. The

Pope was not amused when Galileo wrote a satire about the controversy; it embarrassed the Pope.

1632, Vatican City: "Galileo convicted of heresy by Pope Urban VII for claiming the earth orbits the sun."

The Evidence:

"God fixed the earth upon its foundation, not to be moved forever." Psalm 93:1

"The pillars of the earth are the Lord's and He has set the world upon them." 1 Samuel 2:8

The trial of the most revered of all early scientists set the stage for distrust between science and religion that persists today. It made no difference Galileo declared, "The Holy Bible and the phenomena of nature proceed alike from the divine Word." The Inquisition claimed, "God could produce any observable phenomenon in any way He chose, many explanations are possible, and human interpretations are inherently fallible. Therefore, it is illogical and arrogant to assert a particular example of a physical phenomenon is superior if it contradicts Scripture."

Dominican Father Faccini sermonized against Galileo saying, "Geometry is of the devil and mathematics is the author of all heresies." Besides excommunicating Galileo and placing him under house arrest for the rest of his life, the Church refused to administer Last Rites to him as he lay dying until he recanted.

10/31/1992, CNN: Vatican City: "It's official: The earth orbits the sun. After 360 years, Pope John Paul II declares Galileo not guilty."

The Bible refers many times to the "ends of the earth", which some people still believe means the earth is flat. Even today, most people do not seem to know which way the earth rotates. The globe at Universal Studios and the old TV Daily Show turn the wrong way. For all practical purposes, it is more convenient to think in terms of the sun moving across the sky as opposed to the earth rotating next to it. Hopefully, an educated society will come to realize scientific knowledge is not based on opinions that can be settled by public debate.

The forces of change would result in a tsunami of new approaches and rational thought, the religious establishment, which had dominated

Western society for 1,500 years, would-be put-on defense. For the next four centuries, the very foundations of Western culture would be severely tested by the winds of change. This is most clearly understood by considering the ideas of the people who questioned every assumption upon which our society is built. Theologians would make valiant attempts to justify the Judeo-Christian traditions. Philosophers would make heroic efforts to critically examine the foundation of Greco-Roman culture. But along the way, science begins to move toward center stage. In the end, it now appears the fate of the entire human race is in the hands of scientists. Before we consider the most significant discoveries of modern science, you need to meet the people who shaped the crosscurrents and future direction of the contemporary world.

Sir Francis Bacon (1561 – 1626)

English scientist, philosopher, and statesman, Sir Francis Bacon is credited with establishing the basic principles of the 'scenic method.' The critical logical step of the scientific method is called *a posteriori* (after the fact) or empirical reasoning, as compared to more familiar logical deduction, *a priori* (prior to the fact) reasoning. Science is the systematic study of the properties, culture, and behavior of animate and inanimate in the natural world through observation, experiment, and inference. Basic tenets of the scientific method are (1) we and the universe exist, (2) we and the world are rational, and (3) every natural phenomenon has an explanation which can be discovered by reasonable inquiry. Science is about trusting the scientific method, not the message or the messenger. The scientific methodology is based on repeatable experiments to acquire factual information and logical reasoning, such as causality, induction, inference, prediction, and falsification, followed by peer review, challenge and change.

The first important rule or law of science is "cause and effect" or deduction. The converse of deduction, inference, is the powerhouse reasoning method of science. In this case, reasoning is done backward from the result or effect to the cause. The effect is observed data or empirical facts known to be true, but the cause is not known. A possible reason or cause is called a hypothesis. If all but one explanation or hypothesis can be eliminated, then the assumption or hypothesis is established as a valid scientific theory. The level of confidence of a valid scientific theory must exceed 98%. A scientific theory has the same validity and status as music and art theory. A scientific theory is elevated to a law or fact if it is over 99% certain. Scientific theories and laws are accepted as approximations to some unknowable truth and therefore, ultimately, can be wrong. Scientists are always looking for ways to make arguments better. Sir Francis Bacon wrote, "There are two books laid before us to study, to prevent our falling into error; first, the volume of the Scriptures, which reveal the will of God; then the volume of the creatures, which express his power."

René Descartes (1596 – 1650)

Reconciling reality with our perceptions of reality is an important philosophical issue first considered by René Descartes, who famously concluded in 1630, "Cogito ergo sum," meaning "I think, therefore I exist." With these immortal words, Descartes expressed what he thought was the most fundamental meaning of reality with which thinking persons could agree. He pointed out, "I do not exist," is self-contradictory.

We still want to know if something other than what is in our mind exists. Could existence be explained as a matter of luck? Consider a cosmic lottery. If there was no winner, we could conclude the lottery did not exist. Descartes argues we exist because we won the lottery. Even if we were scammed, we still know we exist because we are here to decide whether or not we got scammed. Unfortunately, explanations based on luck or accidents cannot be used to reason about existence because it is unreasonable to reason about luck or accidents in the abstract. If there is no need for a cause for existence, then there is no need for a cosmic accident to explain it. A cosmic accident is no better as a cause for existence than saying nothing is the cause.

In 1644 Descartes wrote, "I would be quite happy to continue and to show here the whole chain of other truths I had deduced from these first ones. I have always remained firm in my resolve not to suppose any principle, but the one I have just used to demonstrate the existence of God and the soul, and to take nothing to be true does not seem to be clearer and more certain than have the demonstrations of the geometricians been previously. And still I dare say not only I have found the means of satisfying myself in a short time regarding all the main difficulties commonly treated in philosophy, but also, I have noted certain laws God so established in nature and has impressed in our souls such notions of these laws, after having reflected sufficiently, we cannot deny they are strictly adhered to in everything that exists or occurs in the world."

Blasé Pascal (1623 – 1662)

Blaise Pascal, a French mathematician, scientist, and philosopher, devised a pragmatic argument for belief in God. Pascal's wager appeared in Pansies, written in 1654. Pascal maintains we are incapable of knowing whether God exists or not. Reason cannot settle, which way we should favor, but a consideration of the relevant outcomes can:

"God is, or he is not. But to which side shall we incline? Reason can decide nothing here. There is an infinite chaos, which separated us. A game is being played at the extremity of this infinite distance where heads or tails will turn up, which will you choose then? Let us weigh the gain and loss in wagering God is. If you gain, you gain all; if you lose, you lose nothing. Wager, then, without hesitation, He is."

This is the first use of decision theory. There are two types of errors to consider. Type 1 error, false positive, is the mistake of believing in God when He does not exist, and type 2 error, false negative, is the error of not believing in God when He does exist. The best choice is the one with the least penalty for making a mistake. There is no penalty for type 1 error, whereas the penalty for type 2 error could be the loss of eternal life. Therefore, Pascal argues, reason favors belief in God. In biblical terms, "For what shall it profit a man, if he shall gain the whole world, and lose his soul?" Matthew 16:26 The logical implication is clear. Beware of false certainty. The only case where not believing in God is the better choice is proof with absolute certainty God does not exist. It should be obvious such a negative cannot be proved. Pascal would caution, "What can be worse than assuming you know what you don't know!"

Sir Isaac Newton (1642 – 1726)

Isaac Newton was born on Christmas day of the year Galileo died. In 1679, he had an epiphany with an apple. It was a beautiful day, and he was late for a faculty meeting. In 1752 William Stokely wrote about what happened to Newton when he went to the garden and sat beneath an apple tree. "Why should an apple always descend perpendicularly to the ground?", he wondered to himself. "Why should it not go sideways, or upwards? But constantly to the earth's center?" "Assuredly, the reason is the earth draws it; there must be a drawing power in the matter. And the sum of the drawing power in the matter of the earth must be in the earth's center, not on any side of the earth. Therefore, does this apple fall perpendicularly, or toward the center? If matter thus draws matter; it must be in the proportion of its quantity. Therefore, the apple draws the earth, as well as the earth, draws the apple." There is a statue of Isaac Newton, looking at an apple at his feet at Oxford University. It has never been confirmed whether or not the apple hit him on the head.

Newton is the most influential scientist of all time. He was the right person at the right place, at the right time to launch the field of physics in the right direction, while others, like Descartes, had failed. Newton is the father of physics. Every discovery in physics would boil down to a mathematical formula, following the pathway forged by Newton. As an example, Newton's second law of physics states force is equal to mass times acceleration, f = ma. Besides the law of gravity, Newton developed the laws of all dynamic motion, called classical mechanics, the principles of optics, and he shares credit with Leibniz for the development of calculus.

Danish astronomer Tycho Brahe, spent his life making detailed measurements of the motions of the planets. Later in life, he moved to Austria to work on the mathematics of planetary orbits with the brilliant German mathematician, Johannes Kepler. Time ran out on Brahe, so he gave his life's work to Kepler, who used the data to derive the geometric equations of elliptical planetary motion. This enabled Newton to discover the law of gravity. In 1687, Newton published his monumental work, *"Philosophiae Naturalis Mathematica Principia."* Newton derived the geometry of planetary motion using calculus to apply the law of gravity, which he discovered.

Before Newton, no one had even conceived of the possibility causality could occur mysteriously over vast distances with seemingly no physical connection. No one can deny gravity is real, although many have received the Darwin Award for trying. Gravity is a law that must be obeyed. It's the reason we don't fly off a spinning planet. Newton's law of gravity is a simple algebraic equation that describes the trajectory of everything from rocks to rockets. In the orbital motion, two masses rotate around their combined center of gravity or center of mass. Since the sun is a million times more massive than the earth, the center of mass of the earth/sun combination is near the center of the sun. The moon has 1/6 the mass of earth, so astronauts can jump six times higher.

Beginning in 1774, astronomer Sir William Herschel studied the orbits of a double star for nine years to confirm Newton's law was obeyed outside our solar system, establishing gravity is the same everywhere in the cosmos. Gravity works the same if you are standing or jumping. Gravity affects motion, but motion does not affect gravity. Newton's equation applies to all the matter in the universe. It makes no difference what an object is made of, its shape, or its temperature. The object can be a bouncing rubber ball, the ocean tides, a comet made of ice, or a star like our sun made of hydrogen gas burning at temperatures of millions of degrees.

Newton's law of gravity is the single most significant discovery of physical science. It is the foundation of classical physics. It confirms for all time we can comprehend the cosmos in the most powerful and elegant way. It proves beyond any doubt science works. The payoff from science has been enormous. In 1969, a small computer guided the Apollo 11 spacecraft to the moon using Newton's equations.

The force of gravity is the biggest clue to how the entire cosmos works. If you take the time to grasp how gravity works, then the rest of physics will make a lot more sense. Gravity is so crucial in physics a closer look is needed here. You don't need to know anything about calculus to get where we are going. But a little algebra never hurt anybody. Fortunately, the law of gravity is a simple algebraic formula:

$$F = GM1M2/R^2$$

where F is the force of gravity, G is the gravitational constant, and M1M2 is the product is the product of the two masses, and R^2 is the square of the distance between the centers of gravity of the two masses. (The symbol "R" is traditionally used because it was first applied to the "radius" of the earth's orbit.) For purposes here, the pound will be used as the unit of force. Mass is usually measured in kilograms. The gravitational constant, G, in the formula for F, is just a unit conversion factor. The formula is interesting because it is almost intuitive. Notice gravitational attraction is proportional to the product of the two masses. Gravitational force decreases with distance R, just as you would expect. It decreases with the square of distance, R^2 due to the spreading of a two-dimension spherical surface. The force spreads and diminishes spherically in all directions. The surface area of a sphere is proportional to the square of its radius.

Suppose you weigh 150 lbs. This means when you get on a scale, the force of gravity is pulling down on you by that amount. You are motionless because the scale pushes up on you with an equal and opposite force. This is the force, which gets measured by compressing a spring. In physics, it can be said you are motionless because the sum of all the forces on you is zero.

Now consider how much you would weigh if you visited the Spacelab, which is 380 miles above the earth. The diameter of the earth is 3,959 miles, so your weight will be $(3,959/4,339)^2 = 83.25\%$ of your earth weight or 125 lbs. But the Spacelab is circling the earth at 17,500 mph in free space. It is free falling towards the earth under the acceleration of gravity at just the right rate to keep its distance from the earth constant. If it were going too fast, it would go off into space. If it were going too slow, it would fall to earth. It has the Goldilocks speed. Because you and Spacelab are free falling, you can't feel the force of gravity, even though you still weigh 125 lbs. You feel weightless because you and the Spacelab are in free fall together.

Gravity seems strong, but it is a wimp compared to nuclear forces. The nuclear forces inside an atom are 10^{39} (one followed by 39 zeros) times stronger than gravity. A trillion is 10^{12}, so we are talking about forces trillions of times stronger than gravity. (See scientific notation in the Appendix.) The most amazing fact about nature is, it is complicated enough to be exciting and challenging for every generation. But is not so complicated we can't figure it out. This is the Goldilocks nature of the intimate relationship between our world and the human mind. It is like this place was made to give us an interesting puzzle on which to work. On the other hand, there is no scientific reason why gravity even exists at all.

Newton was a devout Christian. He explains the importance of his discovery this way:

"Religion and science are integral parts of the life long quest to understand the universe. God has to be present. He cannot be read out of the universe. This most beautiful system of the sun, planets, and comets could only proceed from the counsel and dominion of an intelligent and powerful Being. This Being governs all things not as the soul of the world, but as Lord over all, and on account of his dominion, He is wont to be called "Lord God" or Universal Ruler. And from his true dominion, it follows the true God is a living, intelligent and powerful Being. He governs all things and knows all things that are or can be done. He endures forever and is everywhere present.

Blind metaphysical necessity, which is undoubtedly the same always and everywhere, could produce no variety of things. All diversity of natural things, which we find suited to different times and places, could arise from nothing but the ideas and will of a Being necessarily existing."

Newton proved the universe is a precise cosmic clockwork mechanism, giving rise to the idea of God the watchmaker. In the eighteenth century, English clergyman William Paley published the first teleological or design argument for God. Paley wrote, "If a pocket watch is found on a heath, it is most reasonable to assume someone dropped it and one or more watchmakers made it, and not by natural forces." The analogy is the complex design and harmony of the universe strongly suggests the work of an Intelligent Creator and not just mindless natural causes."

Gottfried Leibniz (1646 – 1716)

Gottfried Leibniz, was a German philosopher, mathematician, and inventor. He developed calculus independent of Newton, his contemporary in England. The calculus symbology developed by Leibniz is the standard in use today. He also invented the mechanical calculator, the design of which was in everyday use for over a century.

Leibniz developed the ontological argument for God taught in seminaries today. If we are to believe God is the first cause, then, "Who made God?" This age-old riddle threatens to throw reason into an infinite regress because it is a fallacious assumption God is the head of a causal chain. God is not part of a chain; God is the Creator of the chain. Leibniz made this point eloquently by inviting us to consider an infinite collection of books, each one copied from a previous one. To say the content of the books is thereby explained is absurd. We are still justified in asking, 'who is the author?'

The only way out is to assume God can explain himself. This means we can only know God by what He has said and done. This does not mean our finite minds can comprehend God. Saying God is self-explanatory is almost self-evident because such similar concepts arose in all early cultures. The Bible happens to be the most credible self-expression of God.

Logic and reason require it to be established; only God can be the one necessary being. If God were not needed, then the chain of causality could not begin or be attributed to God, and we would want to know what was beyond God, which explains his existence. The concept of a necessary Being always existed or somehow caused his existence has confounded theologians and philosophers since time immemorial. It is indeed an idea that seems to be outside the bounds of logical reasoning. But the fact is, there is no other credible explanation for existence. Our minds are so intricately bound with time, it is difficult to conceive of a life unbounded by time.

The anthropic principle points to an uncountable number of factors upon which human existence is contingent. Gottfried Leibniz wrote, "All contingent things, even if they have existed eternally, lack adequate explanation until they are grounded in something necessary." This is called the argument or principle of sufficient reason. Dutch philosopher Baruch Spinoza developed a similar line of reasoning.

J. L. Mackie argues it is reasonable for something contingent to exist without reason for its existence. This argument defies logic because the term contingent existence means dependence on something else for existence. Religious philosopher Hugo Meynell stated, "God, by his nature, is the sort of being whose understanding and will would explain how it is and it is of everything else without himself being capable of being explained in the same way. On whose existence, understanding, and will, everything else depended could not be dependent, on the existence of anything else.'" What is being said is God has explained his will , but his infinite eternal nature cannot be fathomed by finite temperal beings. On God's existence, understanding, and will, everything else depends and could not be dependent on the existence of anything else. I yield the floor to a non-scientist philosopher and atheist Hugo Parsons, who replied with a cogent point, "There is absolutely nothing in scientific practice or the accepted canons of scientific rationality that entails everything must eventually be explained." So, we must ask, if not God, then what? Scientists who are atheists should at least consent to the fact they have not, and most likely never will find a replacement for or better answer than God. Parsons does not share the view the highest aspiration of science is to know the mind of God, expressed so well by Einstein; "I want to know how God created this world. I want to know his thoughts; the rest are details." [11]

Voltaire (1694 – 1778)

Voltaire was an eighteenth-century French writer, historian, and philosopher famous for his wit, attacks on the Catholic Church, and advocate of freedom of expression and separation of church and state. He was a prolific equal opportunity offender.

Raised as a Catholic, he became disillusioned with organized religion. The 30-Years War between Catholic and Protestant nations left a third of the European population dead and the moral authority of Christianity in shambles. He wrote, "Of all religions, the Christian should, of course, inspire the most tolerance, but until now Christians have been the most intolerant of all men. It is characteristic of fanatics who read the Holy Scriptures to tell themselves: God killed, so I must kill; Abraham lied, Jacob deceived, Rachel stole: so, I must steal, deceive, lie. But wretch you are neither Rachel, nor Jacob, nor Abraham, nor God; you are just a mad fool, and the popes who forbade the reading of the Bible were extremely wise."

In 1763 Voltaire wrote, "It does not require great art, or magnificently trained eloquence, to prove Christians should tolerate each other. I, however, am going further: I say we should regard all men as our brothers. What? The Turk my brother? The Chinaman, my brother? The Jew? The Siam? Yes, without doubt; are we not all children of the same father and creatures of the same God?" He criticized "whites and Christians who proceed to purchase Negroes cheaply, to sell them in America."

French Catholicism and aristocracy were brought down with his wit and wisdom. The French Revolution in 1789 abolished Catholicism. Ten years later Napoleon's coups d'état brought back the Church. But Voltaire helped popularize Deism, which fit the liberal philosophy of his period. Deism rejects the infallibility of the Bible and miracles. Deism replaces religious authority with reason and natural science as sufficient to determine creation requires a Creator. Deism gained prominence among intellectuals in Britain, France, Germany, and the United States.

Examples of Voltaire's wit:

"It is forbidden to kill; therefore, all murderers are punished unless they kill in large numbers and to the sound of trumpets."

"Every man is guilty of all the good he did not do."

"To hold a pen is to be at war."

"God created man, so man returned the compliment."

"If God did not exist, He would have to be invented."

When Voltaire died, Mozart wrote to his father, "The arch-scoundrel Voltaire has finally kicked the bucket."

David Hume (1711 – 1776)

Scottish natural philosopher, David Hume, is considered the first philosopher of the modern scientific era. He wrote prolifically on science and religion. He studied the principles of science using the approach of the immortalized Greek skeptic, Socrates. Hume contended meaningful statements about the cosmos are always qualified by some degree of doubt. He asserts the fallibility of human beings means they cannot obtain absolute certainty except in trivial cases where a statement is true by definition, such as "truth" derived within the framework of a closed logic system.

Thus, philosophical skepticism requires all information must be well supported by empirical evidence. Hume's skepticism does not claim truth is impossible; instead, truth needs the support of "beliefs" derived from other truths. For example, Hume reasoned, "Causal order doesn't follow from logical necessity; it is a fortuitous property of the world, and one for which we can rightly demand some explanation."

In science, knowledge, and truth still exist, just not in all the ways you would think. Critical rationalists like Hume assert scientific theories and any other claims to knowledge can and should be rationally criticized, and if they have empirical content, they can and should be subjected to tests, which may falsify them. They are either falsifiable and thus empirical or not falsifiable and thus non-empirical. This is fundamental to the Socratic Method.

As an example, the naïve empiricism of induction was shown to be illogical by Hume. 1,000 observations of some event do not necessarily make it always true. For example, are all swans white? The goal of science is to prove such hypotheses are true. This seems impossible because it would require us to infer a general rule from some individual cases, which is logically inadmissible. If one black swan is ever found, it must be concluded the statement, "all swans are white" is false. Falsification thus strives for questioning hypotheses instead of proving them. We have a right to question any claim, including religious claims, which can be falsified by empirical evidence.

What if we never knew when the sun would rise or water freeze. Hume found this assumption about nature inexplicable. Why is there natural order exhibited in a world we know was created out of apparent chaos? And moreover, we haven't the slightest rational justification for assuming it will continue tomorrow. If we believe the future will always be like the past when it becomes the past, Hume argues, you are assuming the very thing you are trying to prove. Science can only take the continuing regularity of nature on faith.

As a rationalist and skeptic, Hume made it clear he advocated scientific knowledge over religious beliefs. In his work, "Dialogues Concerning Natural Religion," he presents a debate on the existence of God based on a teleological argument. The character Philo argues against intelligent design because the order in nature may be due to nature alone. If nature contains a principle of order within it, the need for a designer is removed. Philo also argues that even if the cosmos is designed, it is unreasonable to justify the conclusion the designer must be the God of classical monotheism. On the other side of the debate, Hume presents a rational argument for design:

"Look around the world: contemplate the whole and every part of it: You will find it to be nothing but one great machine, subdivided into an infinite number of lesser machines, which again admit of subdivisions, to a degree beyond what human senses and faculties can trace and explain. All these various machines and even their most minute parts are adjusted to each other with an accuracy, which ravishes into admiration of all men, who have ever contemplated them. The curious adapting of means to ends, throughout all nature, resembles exactly, though it much exceeds, the productions of human contrivance; of human designs, thought, wisdom, and intelligence. Since the effects resemble each other, we are led to infer, by all the rules of analogy, the causes also resemble, and the Author of Nature is somewhat similar to the mind of man; though possessed of much larger faculties, proportioned to the grandeur of the work, which He has executed. By this argument some posterior, and by this argument alone, do we prove at once the existence of a Deity, and his similarity to human mind and intelligence?"

If one asks for a rational explanation of existence, then we have no choice but to seek clarification in something beyond the physical world, in something metaphysical because a physical cosmos cannot contain within itself an explanation for itself. What sort of abstract agent might be able to create a universe? It is not sufficient to naively invoke a God instantly creating a world.

Are miracles possible within the framework of science? Hume was only willing to admit scientific evidence as valid knowledge. Unfortunately for Hume, Immanuel Kant points out the very foundation of science, induction, is not real knowledge. Science can only assume experimental results are repeatable and universally valid. Just because the speed of light has been measured 1,000 times and came up with the same answer does not mean it is proved the same answer will result if the measurement is made somewhere else in the cosmos. Referred to as the "black swan" problem, if one is found, the statement "all swans are white" is false. Therefore, Kant concludes it takes faith to believe in unverifiable scientific knowledge.

Equally, so why do we believe in the reality of these? Kant felt reason could remove this skepticism, and he set himself to solving these problems.

Immanuel Kant (1724 – 1804)

"Science is only organized knowledge. Wisdom is organized life." Kant

51

Immanuel Kant was a German theologian and philosopher of metaphysics, epistemology, and ethics. He claimed he had created a Copernican revolution in philosophy. His prolific works provide a fitting transition from the Age of Reason to modern philosophy based on scientific empiricism. His major work, "Critic of Pure Reason," published in 1781, rejects skepticism based on reason by delving deeply into the role of reason in understanding reality.

Kant began with the simple premise; all human knowledge is derived from experience. He wrote, "We first must believe in something to speak at all of knowing and understanding. It is an empirical fact we gain all knowledge by processing signals from the nervous system and our brain, which is stimulated by our sensory perception of the surrounding reality. Kant then asks, "How do we know our perception of reality is actual reality?"

Even Plato recognized the idea of a chair in our mind is different from any particular chair at which we are looking. Also, if we kick the chair, the pain we experience is all in our heads. We don't experience objects themselves; we only experience our mental perception of objects. Called noumenon in the philosophy of Kant, it is the ultimate reality, or "thing in itself," which can be conceived of by thought but cannot be perceived in experience. Our senses are our only means of understanding reality. What then, can we know and not know about reality and existence? The world we know as truth is the material world of phenomena we can know only by empirical science. It is limited to what can be identified by sensory perception. Kant concludes the noumenal world is an invisible reality unknowable by empirical science. Reality exists behind the veil of human experience, just as color is unknowable to the color blind.

In a method similar to Pascal's wager, Kant states critical decision making should proceed as follows: "If one cannot prove a thing is, he may try to prove it is not. And if he succeeds in doing neither, as often occurs, he may still ask whether it is in his interest to accept one or the other of the alternatives hypothetically, from the theoretical or the practical point of view. Hence, the question no longer is as to whether God is a real thing or not a real thing, or as to whether we may not be deceiving ourselves when we adopt the former alternative, but we must act on the supposition of God being real. The presupposition of God, soul, and freedom is then a practical concern, for morality, by itself, constitutes a system, but happiness does not, unless it is distributed in exact proportion to morality, together with life in such a world, which we must consider as future life, or else all moral laws are to be considered as idle dreams."

Thomas Jefferson (1743 – 1826)

During the seventeenth-century, religious wars in Europe, extreme fundamentalist Protestants were targeted for persecution. Puritans, Amish, Quakers, Mennonites, Moravians, and Presbyterians fled to the New World to become the first immigrants in America. Fortunately for them, the Founding Fathers like Thomas Jefferson were highly educated and enlightened men like Voltaire, who firmly believed in religious freedom. While Voltaire was a Deist; Jefferson was a Theist, neither were Christians. Jefferson used scissors to cut the miracles of Jesus out of his personal Bible.

In the Declaration of Independence, Jefferson penned, "We hold these truths to be self-evident, all men are created equal, their Creator endows them with certain unalienable rights among these are life, liberty and the pursuit of happiness." The motto of the 1789 French Revolution reflects Jefferson's words, *Liberty, Élite, Fraternity.*

Jefferson wrote, "Question with boldness even the existence of God, for if there be a God, He must surely prefer honest reason to blindfolded fear. We cannot disregard fact-based scientific evidence and suspend critical thinking to believe in God."

The American Revolution took place during the Romantic period in Europe. Artists rejected the rationalism of the Enlightenment. The artist was free to use his imagination to create original art, "creation from nothingness." Formal traditions were replaced with a more personal relationship with God. Reason took a back seat to individual freedom of expression.

In 1730, Romanticism spawned the evangelical Protestant movement in the American colonies. Called the First Great Awakening, Christianity was promoted as a personal experience, which rejects the ritualism of established denominations. In 1800, the Second Great Awakening swept across America with open-air revivals and missionaries sent to Africa. Even today, Christianity has a message for the whole world. The Bible is a powerful book revered the world over. People of every culture and faith can identify with the insights of each passage. By relating to the spiritual needs of all people, the Bible is a living document. Whatever God is, the Bible gives us access to the living God. America became an evangelical powerhouse as Christianity adapted and assimilated every culture in ea.

In 1830, influential philosophers Ralph Waldo Emerson and Henry David Thoreau promoted a progressive deistic religion in New England called Transcendentalism. It was a Romantic movement based on the ideas of Plato and Kant. They taught ultimate reality is unknowable, nature is divine, and the soul is a spiritual entity that transcends the body and the material world. They reject the need for Jesus as Savior in favor of the inherent goodness of the individual. The Unitarian Church was founded on these precepts. When on tour in Europe, Emerson had to be dragged to visit Voltaire's home, protesting of his unworthiness.

Pierre Laplace 1749 – 1827)

Influential French scholar, Pierre Laplace, chose his profession in spite of his father, who wanted him to be a Catholic priest. Often called the "Newton of France," Laplace was a brilliant mathematician, physicist, and astronomer whose five-volume treatise published in 1825, "Celestial Mechanics," contributed a complete study of all the work of his predecessors with significant advances in the mathematics of physics first developed by Newton. He was the first physicist to realize our solar system emerged from dust and rocks surrounding the sun, much like the rings of Saturn.

Based on his deep understanding of classical physics, Laplace realized if the position and velocity of every particle in the cosmos could be measured simultaneously, then it would be theoretically possible to compute the entire history of the universe backward or forward for all time. This is termed the deterministic cosmos. He opined, "An intelligence, at a given instant, could comprehend all the forces by which matter is animated and the respective situations of the things make it up, if moreover, it were vast enough to submit these data to analysis, would encompass in the same formula the movement of the greatest bodies in the universe and those of the lightest atoms, for such an intelligence nothing would be uncertain and the future, like the past, would be open to its eyes."

From this reasoning, it could then be concluded the cosmos is a mindless deterministic machine churning out all of history. Science would have all the answers. There would be nothing sublime or unknowable about it. Even if there is a Creator, (recall, God is currently resting), Laplace rejected Newton's claim, "God governs all things." According to Laplace, even humans could be fully understood as merely carbon-based molecular machines in terms of biology, biology in terms of chemistry, and chemistry in terms of physics. This is the core belief concerning the nature of reality according to deterministic reductionism of classical physics. Every object in the universe ultimately reduces to a collection of mindless particles. We were made from a concoction of particles, which will disassemble themselves when we die. Nothing more and nothing less is needed. Never again could a scientist be allowed to invoke a need for God in his work. Furthermore, reductive materialism is now used to justify atheism.

When Napoleon asked Laplace why his voluminous work did not mention God, Laplace famously replied, "I have no need of that hypothesis." While Laplace banished God from science, he did not banish God from his life. He never renounced his faith as atheist, Christopher Hitchens claimed. Laplace was a practicing Catholic till death. He kept his religious beliefs private because he did not want them to interfere with his work.

CHAPTER 4

MODERN WORLDVIEW

David Strauss (1808 – 1874)

David Strauss was a German theologian and professor at the University of Zurich who pioneered a critical investigation of historical Jesus. He scandalized Europe with his attack on the core beliefs of Christianity. In 1835, he published, "The life of Jesus", which rejected the divinity of Christ. He claimed the stories about Jesus in the New Testament read like Shakespeare and concluded the Bible was just another library book.

Charles Darwin (1809 – 1882)

Charles Darwin and his wife were members of the Anglican Church. He studied for one year for the ministry before switching to botany. His most beloved daughter, Annie, died of scarlet fever at the age of ten. Darwin would never set foot in a church again.

After his famous trip to the Galapagos Islands on the HMS Beagle, he waited 25 years to publish his work because he knew it would set off a religious firestorm that would lead right up to the steps of the Head of the Anglican Church, Queen Victoria.

Finally, in 1859, the man who would later be recognized as the father of biology, published his life's work, "On the "Origin of Species." In deference to the Church, he placed the following statement at the end of his treatise, "The Creator originally breathed life into a few forms or a single one. From so simple a beginning endless forms most beautiful and most wonderful have been and are being evolved." He also has been quoted, saying, "God designs by laws, not by particular details. All design emerges from evolutionary processes."

A year after his work was published, Thomas Huxley agreed to defend Darwin's work in a debate at Oxford with the greatest orator of the day, Bishop Samuel Wilberforce, who had put a stop to the knighting of Darwin. Wilberforce quickly went for the juggler. He famously asked Huxley, "Were you descended from an ape on your mother's side or father's side?" Huxley replied, "I would rather be descended from an ape than a man who misused his great talents to suppress debate."

What Darwin Didn't Know

Darwin did not invent a scientific theory. His hypothesis was straightforward but incomplete. Darwin's beliefs were verified by discoveries that made his ideas a legitimate scientific fact a century after his death. Darwin didn't know of the astonishing diversity of life on earth. For example, he did not know there are 9,000 bird species, 350,000 beetle species, 250,000 ocean species, 200,000 mammal species known today It is estimated the two million species named since 1758 represent no more than a fraction of the estimated 100 million living world species. Approximately 20,000 new species are discovered every year. 80% of all species live underground. Darwin knew nothing of the brutal chaos and war in nature, which leads to the natural selection of new species. Darwin conceived the idea of evolution, but he had no clue how it worked.

Darwin did not know how evolution worked within interdependent ecological communities. They are not merely controlled by the deadly competition of survival of the fittest. Besides competition, another important part is represented by opportunities for mutually beneficial cooperation or symbiosis, such as between bees and flowers.

Darwin did not know DNA contains the secret of life. The sequencing of human DNA was completed in June, 2000. He did not know DNA contains the Cosmic Code for the information, knowledge, and architecture of all life forms on earth. Bacteria came first and still dominate the planet. He did not know evolutionary convergence has shown eyes and other complex organs have arisen through different evolutionary paths. It is now known life is the product of a logical evolutionary progression of the universe from the beginning of time.

Biologists have shown the same genetic codes are found in all living things on earth: animals, birds, plants, trees, flowers, and algae. We all are related. Humans have 23,000 genes, the same as chickens, an ear of corn, and worms. Some plants have more genes than humans. Darwin did not know embryos form the platform for spawning the diversity of life. He knew nothing about DNA mutation. He did not know 350 million years ago a fish developed fins, which allowed it to go ashore to get away from predators. The Hocking gene modified fins into legs. All four-legged animals have this gene. The bones in the human inner ear look just like the embryonic stage of fish gills. Even though the DNA of humans and chimpanzees differ by only 1%, biologically we differ in over 30 million different ways. The crucial difference is the opposable thumb gene, which eventually supported increased brain size gene mutations.

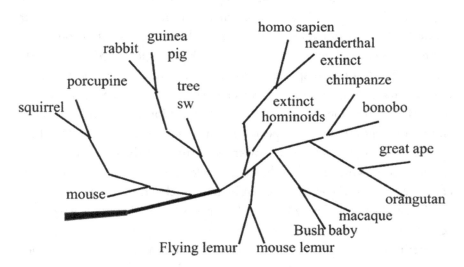

A Branch of the Tree of Life

Darwin knew nothing of the tree of life where scientists use DNA to trace the evolution of all known species down to their roots in single-cell organisms. This chart was created by a computer based on DNA similarities among species. The length of each line represents the closeness of the DNA from one species to the next.

For more than a century after Darwin, natural selection was observable only at the level of whole organisms such as finches or moths, as differences in their survival or reproduction. Now, it can be seen directly how the fittest are made. DNA contains an entirely new and different kind of information than what Darwin could have imagined or hoped for, but which decisively confirms Darwin's concept of evolution. Biologists can now identify specific changes in DNA that have enabled species to adapt to changing environments and to evolve new lifestyles. The new DNA evidence has a significant role beyond illuminating the process of evolution. It could be decisive in the ongoing struggle over the teaching of evolution in schools and the acceptance of evolution in society at large. It is beyond ironic to ask juries to rely on human genetic variation and DNA evidence in determining the life and liberty of suspects, but to neglect or to undermine the teaching of the basic principles upon which such evidence, and all biology, are founded. The antievolution movement has relied on entirely false ideas about genetics, as well as the evolutionary process. The body of new evidence clinches the case for biological evolution as the basis for life's diversity, beyond any reasonable doubt.

9/16/2008, AP, "The Anglican church apologizes to Darwin."

Søren Kierkegaard (1813 – 1855)

Søren Kierkegaard was a Danish theologian and philosopher. A 1964 Life Magazine article recognized him as the father of modern existentialism. Many scholars credit him with being the father of psychology. [12]

Rather than formal abstract philosophy, Kierkegaard concerned himself with the inner human experience. He studied the subjective feelings and behaviors of a real person with all his complexities.

How do people acquire their beliefs, their feelings of anxiety, depression, despair, absurdity, alienation, fear, boredom, pain, guilt, and regret? He would write, "Listen to the cry of a woman in labor at the hour of giving birth, look at the dying man's struggle at his last extremity, and then tell me whether something begins and ends thus could be intended for enjoyment." [13]

Even today, Christianity has a message for the whole world. Kierkegaard recognized the only escape for the individual from social conformity was alienation, feeling alone, confused, and lost. This is Kierkegaard's existential attitude." Only then can you begin to find your true self. Here is a passage in his own words: "What I need is to get clear about what I must do, not what I must know, except insofar as knowledge must precede every act. What matters is to find a purpose, to see what it is God wills I shall do; the crucial thing is to find a truth, which is truth for me, to find the idea for which I am willing to live and die."

Kierkegaard's religious views were intertwined with ethical dilemmas. Abraham had doubted God, so God put him to the test. God commanded him to kill his son. Abraham does not question God's moral authority. Why not? Abraham dared not waver in his trust of God. In return, God no longer wavered in his confidence of the man. God would give his only Son to redeem humankind.

Kierkegaard clearly understood the irresolvable paradox of trying to find common ground between reason and meaning. This led him to the metaphor of a lower story of pessimism ruled by reason and an upper story of optimism devoid of logic. This is the metaphor as explained by Francis Schaeffer: "The optimism of non-reason is absent from the lower story of reason. Downstairs, it is pointless to discuss meaning or values. Reason can't even explain itself. Reason better not try to climb the stairs to the upper story of non-reason where there is optimism and meaning. People who believe they can live only by reason reject the upstairs of non-reason. They try to find meaning in the downstairs of reason. This dichotomy is the existential methodology of Kierkegaard. Non-reason is arbitrary and detached from reason because reason has nothing to offer to non-reason." [14]

In his 1963 book, "The Narrow Pass," George Henry Price wrote: "Kierkegaard was the sanest man of his generation; Kierkegaard was a schizophrenic; Kierkegaard was the greatest Dane....the difficult Dane....

the gloomy Dane; Kierkegaard was the greatest Christian of the century; Kierkegaard's aim was the destruction of the historic Christian faith... He did not attack philosophy as such...He negated reason...He was a voluntarist; Kierkegaard was the Knight of Faith; Kierkegaard never found faith... Kierkegaard possessed the truth; Kierkegaard was one of the damned."

While scientists can learn about the world by observation, Kierkegaard emphatically denied consideration could reveal the inner workings of the spiritual world. He wrote, "If I am capable of grasping God objectively, I do not believe, but precisely because I cannot do this I must believe."

Existential philosophy would lead to the most profound doctrinal differences in modern philosophy, notably distinguished by the presence or absence of God as a precursor to the human condition.

The English writer and philosopher, Gilbert Chesterton, brought the ideas of Kierkegaard into modern psychology. [15] He recognized modern man prides himself on logic and objectivity. Modern man is serious by nature. He does not trust anything which appears illogical. "The only secure truth modern men have is that which they create and dramatize; to live is to play at the meaning of life."

James Clerk Maxwell (1831 –1 879)

Scotsman, James Clerk Maxwell, ranks with Newton and Einstein as one of the three greatest physicists who ever lived. He took up the work left by Danish physicist Hans Ørsted, who discovered electromagnetism in 1820 when he accidentally noticed a compass react to an electric current. Soon after that, English scientist Michael Faraday conducted experiments resulting in a mathematical law, which describes what he measured.

In 1865, Maxwell published his definitive work, "A Dynamical Theory of the Electromagnetic Field." Maxwell formulated the mathematical laws of electromagnetism, EM. In four interconnected equations, Maxwell made the most elegant and powerful use of calculus in history.

The benefit of EM to modern communication is evident in the table. EM is the primary means of long-distance communication in nature as well as between humans. In 1909, Guglielmo Marconi received the Nobel laureate in physics for his invention of the radio communication system, which he used to transmit wireless signals across the Atlantic.

Frequency	Band	10^8	TV	10^{15}	visible light
10^4	very low frequency	10^9	microwaves	10^{16}	ultraviolet
10^5	low frequency	10^{10}	radar	10^{18}	X-ray
10^6	radio	10^{14}	infrared	10^{21}	gamma-ray

Electromagnetic Spectrum (frequency in Hertz)

Light is the supreme informant of the cosmos. Vision is our most important connection to the world. From the farthest star to the book in front of you, no other source can compete with the information about the world provided to us by light.

Light is an EM field, which travels unimpeded through space and time. You can watch TV and have a light on at the same time. Ocean waves need an ocean on which to travel. EM waves do not need a medium to travel wherever they want to go. EM waves can carry devastating power, but for the most part, they are very friendly and useful. Light from the sun carries energy to the earth, which makes life possible.

"Light travels faster than sound. This is why some people appear bright until you hear them speak." Stephen Wright

Frederick Nietzsche (1844 – 1900)

Frederick Nietzsche was a German philosopher and social critic of the late nineteenth century. The son of a Lutheran minister, he came to despise Christianity, which he believed held contemporary culture hostage to absurd beliefs. He derided Christianity because it deprived the evolutionary right of the superior man to subdue his inferior competition. Morality is just a trick to get successful men to give away their hard-earned wealth to losers.

In "Beyond Good and Evil," he traces the development of Western moral systems to explain the central tenet of his philosophy. For Nietzsche, a fundamental shift in social morality took place during human history.

The original form of morality was set by the ruling classes of ancient civilizations, as depicted in Greek literature by Homer. Values of "good" and "bad" coincided with the need for the "master class" to control and dominate the lower classes. He called this "master-slave morality." When Christianity took over Europe, the religious elite would impose the Judaic morality distinguished by "good" and "evil." In other words, humanistic morality was replaced by the objective morality of absolute truth, as revealed in the infallible Word of God.

Declaring the death of God, the atheistic existentialism of Nietzsche is known as nihilism, which is the belief nothing in the world has any real value or importance, and consequently, life has no meaning or purpose. His radical rejection of objective truth is the point of departure for godless philosophy ever since.

Max Planck (1858 – 1947)

German physicist, Max Planck, is the father of quantum physics. His discoveries won him the 1918 Nobel in physics. While analyzing the properties of what is called "black box radiation," he realized the classical equations of Maxwell predicted an infinite answer to the quantity of energy involved. This led him to conclude EM fields are not continuous. Rather, they are constructed from finite increments or units of energy he called quanta. EM quanta, now called photons, carry specific and different amounts of energy, according to the equation, $e = hv$ where energy, e, equals Planck's constant, h, times frequency, v. It took quite a while for Planck to realize the importance of his discovery. He thought of his result as a mathematical anomaly rather than an actual property of reality. For if his discovery were real, it would imply the classical physics of Maxwell was not the whole story. He wrote his formulation was only "a purely formal assumption. I did not think much about it. My unavailing attempts to somehow reintegrate the quantum action into classical theory extended over several years and caused me much trouble." Physicist Max Born would later write, "His belief in the compelling force of logical reasoning from facts was so strong he did not flinch from announcing the most revolutionary idea, whichever has shaken physics."

Planck stated, "Science cannot solve the ultimate mystery of nature.

And is because, in the last analysis, we are a part of the mystery we are trying to solve." Planck was an outspoken religious man. He was a Lutheran who expressed respect for all religious views. He wrote religious "symbols provide an imperfect illustration of divinity." He criticized atheism for its single-minded attack on religious symbols.

In 1944, Planck wrote, "As a man who has devoted his whole life to the most clear-headed science, to the study of matter, I can tell you as a result of my research about atoms this much; there is no matter as such. All matter originates and exists only by a force, which brings the particle of an atom to vibration and holds this most minute solar system of the atom together. We must assume behind this force the existence of a conscious and intelligent mind. This mind is the matrix of all matter."

Planck suggested religious faith should be treated as an unresolved working hypothesis. Faith is a journey, not a destination. He pointed out "causality" is not true or false; it is an act of faith. He said: "Both religion and science require a belief in God. For believers, God is in the beginning, and for physicists, He is at the end of all considerations. to the former, He is the foundation, to the latter, the crown of the edifice of every generalized worldview." On the other hand, Planck wrote, "The belief in miracles must retreat step by step before relentlessly and reliably progressing science, and we cannot doubt sooner or later it must vanish completely."

The implications of modern physics are counterintuitive. You are entering the twilight zone of objective physical reality on scales so vast and so small they deny our practical experience of reality. Forget reality TV. The fact is, you are about to engage in the cosmic mystery theater.

Thompson, Rutherford & Millikin

The Nobel laureate was established in 1905. In 1906, Joseph J. Thompson received the award for his discovery of the electron in 1897. In 1904, he proposed the "plum pudding" model of the atom. The electrons were like raisins in a positively charged gelatin. This is his statement on science and religion: "In the distance tower still higher are scientific peaks, which will yield to those who ascend them, still wider prospects and deeper the feeling whose truth is emphasized by every advance in science, great are the works of the Lord."

In 1899, Ernest Rutherford discovered the alpha, beta, and gamma particles produced naturally by the radioactive decay of the heavy elements. He and Hans Geiger invented the zinc sulfide scintillation screen and the ionization cloud chamber, which can be used to track the motion of atomic particles. Rutherford bombarded gold foil with alpha particles emitted by radium. The scatter pattern of the alpha particles caused by collisions with the gold foil atoms allowed him to measure the size of the atomic nucleus, which turned out to be 1,000 times smaller than the atom itself, which is millions of times smaller than a grain of sand. In 1909, he published the "Rutherford model" of the atom. Rutherford received the Nobel in 1908. In 1917, he would become the first person to split an atom. He is known as the father of nuclear physics. He said, "Of all created comforts, God is the lender; you are the borrower, not the owner."

In 1908, the founder of Caltech, Robert Millikan, measured the electric charge of the electron for which he received the Nobel in 1923.

Bertrand Russell (1872 – 1970)

Bertrand Russell was a British philosopher, logician, and social critic. He particularly admired the work of Leibniz. Privately, he said he was agnostic, but publicly he said he was an atheist. He said he took this position because he could not disprove the Christian God in the same way he could refute the Greek gods.

For most of his adult life, Russell maintained religion was little more than superstition and, despite any positive effects faith might have, he believed it was "largely harmful to people." He believed religion serves to impede knowledge and foster fear and dependency, and is responsible for much of our wars, oppression, and misery. When Russell was asked what he would do if he died and met God, he replied, "I would say, Sir, I wanted to believe in you, but you didn't give me enough evidence." His conclusion, "The only real purpose is the natural continuation of the human species. Nothing is more purposeless, more void of meaning than the world, which science presents for our belief. Amid such a world, if anywhere, our ideals henceforward must find a home. Man is the product of causes, which had no prevision of the end they were achieving, his origin, his growth, his hopes and fears, his loves and his beliefs, are but the outcome of accidental collocations of atoms, no fire, no heroism, no intensity of thought and feeling, can preserve an individual life beyond the grave, all the labors of the ages, all the devotion, all the inspirations, all the noonday brightness of human genius, are destined to extinction in the vast death of the solar system, and the whole temple of man's achievement must inevitably be buried beneath the debris of a universe in ruins, all these things, if not quite beyond dispute, are yet so nearly certain, no philosophy, which rejects them can hope to stand. Only within the scaffolding of these truths, only on the firm foundation of unyielding despair, can the soul's habitation henceforth be safely built."

Albert Einstein (1879 – 1955)

"My religiosity consists in humble admiration of the infinitely superior spirit that reveals itself in the little we, with our weak and transitory understanding, can comprehend of reality." [16] Einstein.

Einstein is the definition of genius. He was the quintessential physicist's physicist. The biggest question left in physics today is; who will receive Einstein's mantle?

In 1905, at age 26, Albert Einstein came out from behind a desk at the Swiss Patent Office to become an instant celebrity in the world of physics. In one of his five seminal papers of that year, he developed the most famous equation in science, $E = mc^2$. In 1921, he received the Nobel at age 32. In 1930, he came to Caltech as a visiting professor. As World War II loomed, he helped over 1,000 Jewish scientists escape from Germany. Winston Churchill said the Nazis had "lowered their technical standards, to the benefit of the Allies."

Refuting atheism, Einstein said he believed in the God of Spinoza.

While Newton's gravity is still the law of the cosmos, Einstein would discover it needed a significant modification to conform to the work of Maxwell. Modern physics now rests on two pillars, Einstein's law describes the universe on the grandest scale, and quantum mechanics describes the universe on the smallest scale. These two theories must come together to complete the physical laws of the cosmos.

Applying the principle of conservation of mass and energy, he discovered the basic mechanism of the atomic bomb. Matter can be turned into energy and energy, can be turned into matter. He proved mass and energy are equivalent according to the equation, $E = mc^2$, where "c" is the scientific symbol for the speed of light. On 8/6/1945, one third of an ounce of uranium was transformed into energy equal to 15 kilotons of TNT at Hiroshima, Japan.

A discussion of Einstein's theory of relativity begins with light. Descartes reasoned logic requires the speed of light to be infinite. In 1676, Danish astronomer, Ole Roemer discovered it was finite quite by accident. Galileo had suggested the orbital periods of Jupiter's moons could be used as celestial clocks for navigation. While studying these orbits, Roemer noticed the time at which the moons were eclipsed by Jupiter varied over a year. The time between eclipses was most significant when the earth was at its furthest distance from Jupiter. This could only mean the time it took light to reach earth depended on distance. The speed of light was finite. Light sets the speed limit of the cosmos. If light traveled instantaneously, we would be locked in time and astronomers would not be able to study the history of the cosmos. We would never know how we got here.

In 1887, Albert Michelson made very accurate measurements of the speed of light by calculating the time it took for light to travel between Mount Wilson, Mount Baldy, Mount San Jacinto, and Mount San Gorgonio, distances from 18 to 22 miles. There is a stone monument near Idyllwild, CA, commemorating this important experiment for which Michelson received the 1907 Nobel. Scientists finally realized the speed of light was a universal, absolute cosmic constant of fundamental significance to understanding the universe. Stephen Wright weighed in, "OK, so what's the speed of dark?"

At the time of Michelson's experiment, it was believed there must be a medium in space to support EM wave propagation through it. The invisible substance was called "Aether." Michelson conducted experiments to measure the speed of the earth through the Aether. By measuring light speed with and against the motion of the earth, he could factor out the earth's speed from the difference in the two measurements. The result was zero. Either the earth does not move, or there is no Aether. There is no Aether. Light travels through the vacuum of space without the need for a medium to support it. This created a new intractable problem for a genius to solve.

Speed is merely a measurement of distance divided by time, $s = d/t$. Because there is no frame of reference in space, the concept of speed seems somewhat nebulous. Nevertheless, speed in space is intricately tied to the fact light is traveling through it. It took Einstein to figure out how to understand the concept of speed in outer space.

Einstein thought about the problem while riding his bicycle to the Swiss Patent Office. How would he know how fast he was going if there was no ground beneath his bike? Indeed, there would be no way to know. But he knew he could at least know how fast he was going relative to other bicycles and motorcars. This is the idea of "relative speed," the speed of two moving objects relative to each other.

If you were driving a car at a constant speed on a freeway, there is no sensation of motion. According to special relativity, you would be justified in claiming you were stationary and the rest of the world was moving by you. Only the ground around you provides a frame of reference that convinces you this is not the case. So, Einstein realized in outer space; it was just the relative speed between objects that could be measured. And in the fact of Maxwell's EM theory, this relative speed is a universal constant, the speed of light.

Furthermore, the speed of light was an absolute barrier to relative motion. The relative motion of any two objects cannot exceed the speed of light. In classical physics, two objects traveling toward each other, each at the speed of light, would have a relative speed of twice the speed of light. Only Einstein could solve this intractable dilemma. These properties of light led Einstein to discover the theory of special relativity.

Einstein's theory was revolutionary because it defines the speed of light as a fundamental constant or law of nature. This has the remarkable implication all measurements of speed, distance and time must comply with this law of the cosmos. Fortunately, the speed limit of the cosmos is high enough so we won't be needing space cops.

At this juncture, it will become apparent there are a lot more mysteries underlying reality than quantum mechanics. Welcome to the twilight zone where time loses its meaning. Einstein's theory concludes time is different everywhere in the universe. Time is indelibly tied to the fabric of space.

As space expands, so does time. We cannot directly experience the expansion of space and time because it is an effect, which is most noticeable over cosmic distances and times. In particular, time slows down relative to time on earth for objects such as the furthermost galaxies, which recede from the earth near the speed of light. Scientists call this effect time dilation.

To understand what is going on, consider an extreme Gedanken or thought scenario where Albert has a bud, Buzz Lightyear, who sends a message that he just left for earth from Planet X in another star system, which is one light-year from earth. Albert knows Buzz went to Planet X one year earlier than when he received the message because it is how long it takes for light to get to earth from Planet X.

Meanwhile, Buzz is in his spaceship, traveling to earth at light speed. This means messages sent to Earth by Buzz are traveling towards earth at the same speed as Buzz in his ship, and therefore the signals will arrive at earth at the same time as Buzz.

Stephen Wright once asked Einstein a related question; "If your car could travel at the speed of light, would your headlights do you any good?" Einstein said they would work just fine because light speed is relative to your speed. Light doesn't know or care about your speed; it only that cares its speed relative to you is the universal constant.

At first, nothing seems violated here because the speed of Buzz's spaceship is relative to earth and light signals from Buzz's spaceship know nothing about the earth, except their speed relative to earth must be the same universal constant. NO! Something is wrong.

Albert knows Buzz's messages cannot travel to earth faster than Buzz's spaceship. But Buzz knows the messages must be traveling toward earth at the speed of light on top of the speed of his spaceship. If the messages are traveling at the speed of light relative to Buzz, then they must be traveling at twice the speed of light relative to earth. Sorry, this is against the law.

Einstein knows Buzz and Albert are both right. Messages from Buzz travel to earth at the speed of light relative to Buzz, and they arrive at earth at the speed of light relative to earth. Einstein realized the discrepancy could only be explained by allowing clocks to run at different speeds depending on the relative speed of their locations. This is the only way to preserve the constancy of light speed. If this doesn't make sense to you, don't lose sleep over it. I still do too. Just believe it because it's the law!

Here is what you have to believe. The solution is time has stopped on the spaceship relative to time on earth. When Buzz gets to earth, he will be the same age as when he left Planet X. This does not mean Buzz was in some suspended animation. Life on the spaceship was perfectly normal for Buzz. He didn't get younger either. What happened is time slowed to a standstill for Buzz only relative to time on earth.

If Buzz had been traveling at light speed from a star which is a million light-years from earth, a million years would have passed on earth yet he would not have aged a bit. If Buzz had a twin brother who lived on earth while Buzz spent a million years traveling at light speed, when Buzz returned to earth his brother would have been dead for a million years while Buzz would be the same age as when he left. This is not a gimmick or scientific sleight of hand. It is reality. When you or an astronomer looks at the light from an object, no matter how near or far, you are seeing the light exactly as it appeared when it left the object because, for the light, no time had passed.

This mystical phenomenon is real and defies everything associated with Newtonian physics. There is nothing intuitive about it because we move about at a snail's pace compared to the speed of light. To accept special relativity is an intellectual commitment to abandon the commonsense notions of Newton's absolute space and absolute time.

The strange fact of special relativity has within it a possible explanation for why Genesis states God created the universe in six-days. Given we have been traveling away from the big bang near the speed of light for 13.8 billion years, what can be said about the Creator's view of time if he appeared to someone on earth to explain his creation? Time dilation could explain the premise of Genesis. The answer is little or no time would have elapsed for the Creator. Perhaps six-days of events is a real and literal explanation of creation from the Creator himself, who after all, is directly credited with telling the story to some anonymous listener.

I have a hunch the reality being discovered by science is far stranger than the mysterious stories in the Bible. It is possible the reality we experience is nothing more than a fiction of our imaginations, which we must believe in order to exist.

Newton's world ran by a master clock. This is a core assumption of classical physics, which stood for centuries until Einstein entered the scene. The structure of time is built into our daily lives. As natural as this way of thinking may seem, you will not find time in modern physics. Einstein's theory of relativity treats all moments in time as equally real. Fundamentally, the present and the future are no different than the past. The difference between the scientific understanding of time and our everyday experience of time is troubling, to say the least.

The idea of a timeless reality is initially so startling it is hard to see how it could be coherent. We now know reality is the totality of all the events in space and time. Just as we envision all of space as really being out there, as really existing, we need to accept all of time is out there too. Past, present, and future appear to us as distinct entities. The only thing that is real is the whole entity of space and time, now a four-dimensional object called, spacetime in physics. Einstein notes, "The distinction between past, present, and future is only an illusion, however persistent." "For us believing physicists, the demarcation between past, present and future has merely the significance of but a persistent illusion." [17] Physicist John Wheeler put it succinctly, "Time is nature's way to keep everything from happening all at once." Likewise, space keeps everything from happening in one place.

Conscious experience seems to move through time as though our minds' perception is a projector, so moments in time come to life when the power of consciousness illuminates them, the flowing sensation from one moment to the next.

In 1915, after ten years of theoretical research, Einstein made his most significant discovery, an equation that integrated electromagnetism, gravity, and spacetime into a single coherent theory called general relativity. Einstein's theory describes the geometry of the spacetime field as the motions of matter dynamically warp the shape of space and time.

$$R_{\mu v} - \frac{1}{2}g_{\mu v}\,R + g_{\mu v}\,\Lambda = \frac{8\pi G}{c^4}T_{\mu v}$$

where R is the Riemann complex manifold curvature tensor transformation of spacetime, expressed by a non-linear partial second order differential vector matrix with four time-dependent variables. (Einstein speak) Λ is the cosmological constant, G is Newton's gravitational constant and c is the speed of light. Again, he based his theory on complex mathematical formulas and fantastic insight. Newton's first law of physics states any object moving through space in a straight line at constant speed experiences no forces. If the trajectory of an object follows a path other than a straight line at constant speed, it is accelerating under a force, according to Newton's second law. The geometry of a trajectory in space determines whether an object is accelerating or not. But in Einstein's theory of general relativity, space and time, not space alone, are the ultimate arbiters of accelerated motion.

Accelerated motion is different from constant motion because you can feel it. You feel it in a car when you step on the gas, hit the brakes, or make a quick turn. Einstein realized the similarity between accelerated motion and gravity was not a coincidence. With incredible insight, Einstein realized gravity and accelerated motion are identical in effect. Einstein realized if you found yourself falling under the force of gravity, the acceleration you would experience is the same as the acceleration you would experience in a car applying the same amount of mechanical force, called one G, directly quarter-mile car race at one G would take 7.5 seconds with a top speed of 240 mph. Einstein's most significant insight is gravity must obey the speed of light. In one fell swoop, he brought Newton's theory of gravity into compliance with Maxwell's theory of electromagnetism. Einstein realized gravitational force is not a direct force on matter, as thought by Newton. He discovered gravitational force is caused by matter changing the shape of spacetime. The force of gravity warps space into a concave funnel shape into which objects fall. Think of a large object such as the earth sitting at the bottom of a spacetime depression or funnel.

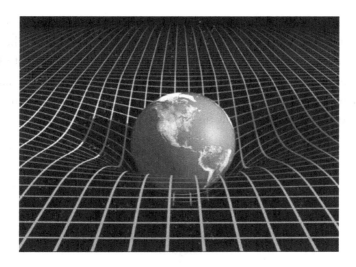

Gravitational warping of spacetime

In general relativity, space and time are not just there. They are movers and shakers in the cosmic mystery theater. Spacetime is like a fabric made from a rubber membrane. Matter causes a depression in space, which causes other matter to move, which causes depressions in space to interact, and so on. [18] This is how gravity warps the fabric of spacetime. Greater mass causes greater depressions. General relativity is a four-ring cosmic circus of space, time, energy, and matter.

Einstein's theory found gravity and Electromagnetism (EM) are closely related. As objects move, the gravitational fields move in spacetime like waves traveling at the speed of light. If the sun suddenly disappeared, it would take eight minutes for the earth to know it was supposed to stop revolving around where the sun had been. Gravitational fields bend the path of light through space. In 1919 this phenomenon was confirmed by astronomer Sir Arthur Eddington. He followed the path of light from a distant star as it was bent by the sun during a solar eclipse.

Einstein's greatest theories are the result of his imagination and mathematical genius. His ideas were far beyond scientific thinking of his time. Experimental proof only followed much later when other scientists figured out how to test them.

Einstein wrote, "What is so incomprehensible about the universe is it appears to be comprehensible. There is a spirit manifest in the laws of the universe. In the face of this spirit, we have to be awed and humbled. The beauty of the universe should humble us. The awe of the cosmos is a source of religious inspiration but challenging to elucidate. I want to know how God created this world. I am not interested in this or that phenomenon in the spectrum of this or that element. I want to know his thoughts; the rest are details." [11]

Niels Bohr (1885 – 1962)

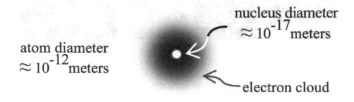

Bohr Model of the atom

In 1912, Danish physicist, Niels Bohr, developed the first scientifically accurate model of the hydrogen atom, for which he received the Nobel in 1922. The "Bohr model" modified the Rutherford model by incorporating the ideas of Max Planck. The atom is still represented by the planetary analogy, but the orbits of the electrons are constrained to fixed quantum energy states called shells. Once the lowest energy state is measured, e_0, then the higher energy states are determined incrementally by the formula, $en = e0(1+1/n^2)$, where $n > 1$.

Bohr proved Planck's quantum EM theory also applied to matter. In the Bohr model, the electrons normally occupy their lowest energy state, called the ground state. The ground energy state cannot be zero, otherwise the EM attraction between the negative electron and the positive protons of the nucleus would cause the atom to collapse. The electron can jump to a higher state or shell if it is excited (i.e. given energy) by some outside energy source such as a photon.

The Bohr model needs some tweaking to accurately describe the heavier atomic elements. Credit goes to Wolfgang Pauli, who developed the mathematical theory of quantum spin in 1927, for which he received the Nobel in 1938. The spin of an elementary particle is an intrinsic quantum energy property, just like the electrical quantum charge and the quantum mass-energy. Spinning is rotational energy, like an ice skater converts forward motion energy into spinning energy. A particle's spin describes its magnetic field, just like spinning produces the earth's magnetic field. Pauli's exclusion principle states no two electrons can occupy the same spin state inside an atom. The quantum properties of electrons in an atom form the organizing principle of the periodic table, which is the foundation of chemistry. Electrons are the workhorse of the cosmos. They are responsible for bonding atoms together, making every kind of molecule, from which all matter is made. The changing of the quantum state of an electron either creates a photon or absorbs the energy of a photon, across the entire EM spectrum. This mechanism creates all the light in the universe, from stars to campfires.

Bohr's double slit experiment has baffled physicists to this day. But the experiment is real; the results are real; the facts are real. The experiment has been reproduced thousands of times by researchers and physics students. The experiment has been tried in every conceivable way in order to figure out what is happening. It is the single most important scientific experiment in history. It would change the scientific view of reality forever. Victorian era physics, which reduced matter to atoms, was just a stepping stone along the path towards a deeper understanding of reality, perhaps even the bedrock foundation of reality. Quantum physics is the new sheriff in town. Quantum physics is the ultimate enforcer of the laws of the cosmos.

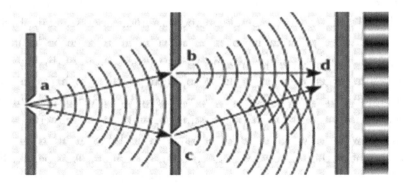

Double Slit Experiment

The effect is analogous to a wave on the ocean passing through two small inlets in a seawall, which protects a harbor. As the wave passes through the inlet channels, the wave acts as two waves with patterns identical to the original wave coming into the harbor. The two waves have the same frequency or period, which is the distance between the peaks and valleys. The critical point is for small inlets comparable to the period of the waves, the two wave patterns spread out as they enter the harbor. The peaks and valleys of the two waves interact constructively and destructively to form what is called an interference pattern, or diffraction pattern, as shown on the right in the diagram. Notice how dramatically different this result is compared with what happens as two boats enter the harbor. The two boats do not need to interfere with each other because they have separate paths. They can both enter the port at the same time without fear of colliding, even though their bow waves do interfere with each other.

The quantum principle is that two particles, each passing simultaneously through adjacent narrow slits, do not act like particles; they produce the interference pattern of two waves on the other side! This result has been observed using two molecules with more than 100 atoms. Now here is the most bizarre result of all. If a single particle is fired at the two slits, it will interfere with itself as though it was a wave that passed through both slits! The result is called the complementary principle. All quantum particles exhibit the dual wave-particle property of quantum mechanics. This fact applies to every particle in the universe. It does not matter if it is an EM particle with no mass, a photon, or any of the many particles with mass.

Now I shall tell you the best kept, and least understood secret in physics. The double slit experiment has been tested with detection sensors located at the entrance and exit of the two slits to get to the bottom of this mystery. The result turned out to be a philosophical puzzle, which is the most confounding discovery of science ever witnessed by a scientific instrument or a conscious mind. The particle passing through the experimental setup "knows" when a conscious mind is watching it. Instruments record the experiment as though a wave went through the apparatus as before. But if a human monitors the instrumentation, the particle passes through the apparatus just exactly as a particle would be expected to do in classical physics. This is called the "observer effect." It violates a fundamental principle of science established by Galileo; experimental results must be independent of the observer.

At the conference in Copenhagen, Bohr presented his findings. Erwin Schrödinger recalls sitting on a bench one night, pondering the experiment, when he had an accidental epiphany. Across the street, a man was walking along the sidewalk. He could only see the man when he walked under a street lamp. It occurred to Schrödinger that he believed the man was there when he could not see him, but he could not prove it. He realized Bohr had proved a particle literally becomes an invisible wave when no one is looking at it!

It is spooky to say the least, because it gives researchers the impression quantum particles have an awareness of the presence of conscious minds. One thing is for sure; it demonstrates the fundamental limitation of the ability of the observer to predict experimental results. In 1963, Nobel physicist, Eugene Wigner, explained the observer effect this way, "It is not possible to formulate the laws in a fully consistent way without reference to consciousness. Consciousness gives an account of its state by causing a collapse of the wave function into a particular value. This does not mean consciousness creates reality. It means the act of communication with the quantum world results in the transfer of information to an observer's consciousness, which results in the collapse of the wave function."

Richard Feynman called the double slit experiment "the heart of quantum mechanics. In reality, it contains the only mystery of quantum mechanics."

The bottom line of quantum mechanics is that all particles, when isolated from all other particles, are standing waves similar to the vibrations of a violin string as shown in the diagram below. Better yet, quantum waves are more like the three-dimensional vibrations of a drum. The amplitude or envelope of the quantum wave indicates the likelihood or probability of where the particle effect may occur at any given time. The amplitude of an isolated quantum wave has the shape of the wave of a rock thrown into a pond.

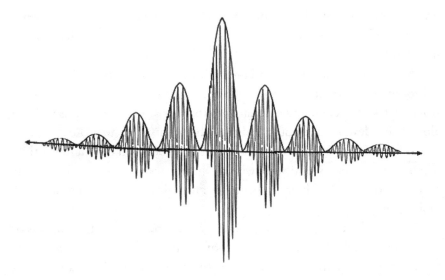

Free Standing Quantum Wave

The precise location of a particle can never be known in theory or measured in practice. It makes no difference whether it is a massless photon or an atomic particle. The center of the probability shape is the most likely place to find the particle effect. The probable particle location decreases with distance from its most likely location, but never reaches zero. This means an experimenter can predict where a particle should be, but when he tries to observe its location, it can be anywhere in the universe. Even more bizarre, though the "particle effect" can be observed, atomic particles are not really particles. An atomic particle is a standing wave with the same geometric shape as the probability distribution of its "apparent location." What seems like a particle is nothing more than a ghostly wave!

The French nobleman and physicist Louis De Broglie received the 1929 Nobel for his 1924 17-page PhD. thesis, the shortest in the history of physics. De Broglie discovered the formula from which the wavelength of a particle, λ, (i.e. Greek lambda) is determined from its momentum, p, by the equation,

$$\lambda = \hbar/\text{p},$$

where wavelength, lambda, equals Planck's constant, h bar, divided by momentum, p, where p is mass times velocity, p = mv. Notice, a stationary particle with p = 0 is not allowed in quantum physics.

De Broglie's theory was experimentally confirmed in 1927 by two American physicists. Physicists now know matter is not what it seems. While matter is made of molecules, which are made of atoms, which are 99.9999999% empty space. The only thing there is fuzzy nuclei surrounded by a fog of electrons. The electrons form a shield protecting the nuclei from collisions with each other. British physicist Sir Arthur Eddington said, "Matter is mostly ghostly space." If you think quantum physics is hard to understand, join the club. Feynman wrote, "I think I can safely say nobody understands quantum physics." I would dare say the Creator of quantum physics knew precisely what He was doing. The quantum layer is the impenetrable floor of reality, a permanent barrier to the prying eyes of science.

The famous Bohr-Einstein debates took place at the Congress of Solvay in late 1920s. They were examining the deepest philosophical foundations of science. Bohr defended quantum mechanics as he had established from his double slit experiments. Bohr prevailed in what was a contest of wills and reason more than science. Bohr published the points of dispute in an article, "Discussions with Einstein on Epistemological Problems in Atomic Physics."

It could be said Einstein was too old school to accept Bohr's interpretation of the double slit experiment. Einstein argued Bohr's theories were incomplete because they did not have a foundation in the tried-and-true principles of science, particularly causality. Indeterminism had never before been a part of physics. Newton and Maxwell had proved the universe runs according to causal laws, with no exceptions. At the core of classical physics was the belief the universe is a well-ordered machine. To Einstein, quantum theory was heresy. Einstein rejected quantum physics because it was incomplete and incomprehensible. Einstein famously told Bohr, "God does not play dice with the universe." Bohr replied, "Who is Einstein to tell the Lord what to do."

Einstein never saw the need to go inside a laboratory. His laboratory was in his head. He is famous among physicists for conducting "bedarken" or thought experiments. In the Bohr to Einstein debates, Einstein challenged

Bohr's experimental methods with his own mental versions of the double slit experiment. Einstein attempted to establish microscopic experiments cannot be trusted when they are conducted with macroscopic instruments. It wasn't easy, but Bohr won the argument. It turns out the philosophy of physics comes down to picking the right equations to describe reality. For any given phenomenon, only one theory, no matter what it is or who discovered it, only one can be the right one.

These exchanges of ideas would be critically important to the future of nuclear physics. Physicists were already experimenting with atomic chain reaction, which would make nuclear reactors and atomic bombs a reality.

Werner Heisenberg was a German physicist who closely followed Bohr's work. He won the 1932 Nobel for his 1905 mathematical theory of indeterminism, the Heisenberg uncertainty principle. It states the measurement error of the location of a particle, times the measurement error of the momentum of the particle, cannot be less than 1/2 Planck's constant.

$$e(x)e(p) > \hbar/2$$

He would soon become head of research in Germany with the goal of making the first atomic bomb.

Bohr told his students, "Reality is stranger than science fiction." In 1909, Bohr gave his brother a book by Kierkegaard with the note, "It is the only thing I have to send home, but I do not believe it would be straightforward to find anything better. I even think it is one of the most delightful things I have ever read." But Bohr mentioned he disagreed with Kierkegaard's philosophy. Bohr valued reason too highly to allow faith to cloud his thinking. He never openly discussed religion, but he was most likely a Theist like Jefferson.

Paul Tillich (1886 – 1965)

Paul Tillich, the founder of the Harvard Divinity School, was an influential proponent of liberal Christianity. His belief system is a form of secular philosophy or existential humanism. He considered Jesus a great teacher but rejected his claim of divinity. His primary interest was the interpretation of Christian symbolism. Like Kierkegaard before him, he concluded there is no rational explanation for God or existence. God cannot be derived from human existence. God can only be known through revelation. But belief in Scripture must be predicated on the answers it gives to questions of human existence. And faith cannot conflict with a rational understanding of human existence. The problem for Tillich is the practice of Christianity has reduced God to a few words with no meaningful content. In order to explain God, Tillich proposes an ontological argument for God along the lines of Anselm, who 800 years earlier considered the meaning of "being" to explain God. In other words, philosophy always ends up being the study of what it means to exist.

For Tillich, the ontological question must be answered by an independent analysis of humans and God as beings to determine their relationship to each other. For this purpose, he accepts a priori the God whose Word became flesh, Jesus, is the same as the logos of Greek philosophy, the "Unchanging One." [19] (See the Unknown God speech by Apostle Paul in Acts 17:23) Tillich finds the answer to the theological and philosophical question of existence is to define God as the "ground and power of being." Man's existence is being as opposed to non-being, is then contingent on the necessary existence of God. But God is not a being. All we can say is God is just another word for existence.

Francis Schaeffer, American evangelical Christian theologian and philosopher, had this to say about existential liberal theology in general and Tillich in particular; "They do not accept God in the Bible, and the revelation in Christ have given man truth, which may be expressed in logical propositions. For them, all content about God is dead, and all assurances of a personal God is dead. One is left with the emotion of certain religious words and that is all." [14] Freedom of religion in America resulted in many "modernized" versions of Christianity in the late nineteenth century, thereby throwing the baby Jesus out with the proverbial bathwater. Liberal Christianity is like Eastern mysticism; the teachings of Jesus and Buddha are both guides to a good life. They celebrate life through prose, poetry, and music. God has no role in human affairs nor the vagaries of nature. God has not revealed himself in any way. He does not listen to nor answer prayers. Religion is just a human invention.

By denying the divinity of Jesus, liberal forms of Christianity gut the Bible to broaden its appeal to those seeking moral guidance, a feel-good philosophy of love and compassion without the baggage of a real personal God. "It's a user-friendly Jesus who doesn't make demands on someone." Ben Witherington, New Testament scholar.

Erwin Schrödinger (1887 – 1967)

$$i\hbar\frac{\partial}{\partial t}\Psi(\mathbf{r},t) = -\frac{\hbar^2}{2m}\nabla^2\Psi(\mathbf{r},t) + V(\mathbf{r})\,\Psi(\mathbf{r},t)$$

Schrödinger's equation

Erwin Schrödinger formulated the quantum wave equation in 1925, for which he received the 1933 Nobel. His equation for quantum mechanics is no less fundamental in physics than Einstein's equations of relativity. An atomic particle trapped within a molecular structure is described by a wave equation which is called a harmonic oscillator. It can be visualized as the motion of a child on a swing, hence the term, oscillator. Only harmonically related frequencies of the waves are allowed. The wave is trapped inside an energy well. Even more bizarre, an atomic particle can escape the oscillator when least expected. Just as a child might try to jump off a moving swing. This phenomenon, called quantum tunneling, is an indispensable part of microelectronics because it describes how electrons move about in a semiconductor.

The philosophical problem of "Schrödinger's cat" endures as his legacy in popular science. A quantum particle can be in more than one place at the same time. He posted a thought experiment to Einstein to illustrate the point. He imagines placing his cat in a box with a quantum particle, which will release a vial of poison if it moves. Since the particle can stay in the same place and move somewhere else at the same time, he claims the cat must be assumed to be both dead and alive until he looks in the box, causing the wave function to collapse and forcing the particle to make up its mind. In spite of the skepticism of Einstein and others, Schrödinger stated, "The mathematical framework of quantum theory has passed countless successful tests and is now universally accepted as a consistent and accurate description of all atomic phenomena."

Schrödinger said, "Physicists have not been able to provide a metaphysical interpretation of quantum physics." He looked to Plato, Kant, and Einstein to study epistemology and search for answers to the mystery of life. Quoting Plato, he wrote, "A mathematical truth is timeless, it does not come into being when we discover it." He agreed with Kant, who believed time and space exist only as constructs to satisfy the mind. Schrödinger wrote, "Einstein has not, as you sometimes hear, given the lie to Kant's deep thoughts on the idealization of space and time; he has, on the contrary, made a step towards its accomplishment."

Schrödinger decided his work was a metaphor for the philosophy of Buddha. He believed consciousness is our universal connection to the cosmos. Schrödinger declared the mind is the "elephant in science." The universe only exists in mind. He said physicists pretend mind is not a part of their work. Concerning Bohr's observer effect, he wrote, "A careful analysis of the process of observation in atomic physics has shown the subatomic particles have no meaning as isolated entities, but can only be understood as interconnections between the preparation of an experiment and the subsequent measurement." He adds, "The world is given to me only once, not one existing and one perceived. Subject and object are only one. The barrier between them cannot be said to have broken down as a result of recent experience in the physical sciences, for this barrier does not exist. Quantum physics thus reveals a basic oneness of the universe." He concludes there is only one universal mind, which is made manifest by the human mind. He states there would be no reason for the universe to exist without a conscious mind. He wrote, "What we observe as material bodies and forces are nothing but shapes and variations in the structure of space. When God is experienced... He must be missing in the spacetime picture just like our minds are." He believed the universal mind is the immaterial presence, which is the essence of existence itself. He wrote, "To me, the contrived nature of physical existence is just too fantastic for me to take on board as simply "given." Some of my colleagues embrace the same scientific facts as I, but deny any deeper significance. They shrug aside the breathtaking ingenuity of the laws of physics, the extraordinary felicity of nature, and the surprising intelligibility of the physical world, accepting these things as a package of marvels just happen to be. But I cannot do this."

"We do not belong to this material world that science constructs for us. We are not in it; we are outside. We are only spectators.

I am very astonished that the scientific picture of the real world around me is deficient. It gives a lot of factual information, puts all our experience in a magnificently consistent order, but it is ghastly silent about all and sundry that is really near to our heart, that really matters to us. It cannot tell us a word about red and blue, bitter and sweet, physical pain and physical delight; it knows nothing of beautiful and ugly, good or bad, God and eternity.

Sensations and thoughts do not belong to the "world of energy.". The reason why we believe that we are in it, that we belong to the picture, is that our bodies are in the picture. Our bodies belong to it.

Mind and Matter." Schrodinger chapter 5, Institute for Advanced Study in Dublin, 1958

Clive Staples Lewis (1898 – 1963)

Professor of literature at Oxford, C. S. Lewis was raised as a Catholic in Belfast, but he found church boring and the strife between Catholics and Protestants turned him to atheism at 15. He quoted Lucretius as the best known argument for atheism, "Had God designed the world; it would not be a world so frail and faulty as we see." In adulthood, he would say he was "angry with God for not existing" At Oxford his friend J. R. R. Tolkien of "Lord of the Rings" fame, dragged him to Protestant church.

Lewis is well known for his writings, "The Chronicles of Narnia" became a hit Hollywood movie. His best-known Christian work is "Mere Christianity." [3] Like Spinoza, he realized it is paramount God be understood as a rational Being. He brought an ecumenical perspective to Christianity, which is widely accepted by Catholics and Protestants. His book "God in the Dock" emphasizes the importance of reason in justifying faith. It brought Christian apologetics out of the staid theological woodwork. He also made an important contribution by bringing modern science into his interpretation of the Bible. He has been dubbed "Apostle to Skeptics."

Here is considered the most crucial passage in all of his writings: "I am trying here to prevent anyone from saying the foolish thing people often say about Him: I'm ready to accept Jesus as a great moral teacher, but I don't take his claim to be God. It is one thing we must not say. A man who was merely a man and said the sort of things Jesus said would not be a great moral teacher. He would either be a lunatic on a level with a man who says he is a poached egg, or else he would be the devil of hell. You must make your choice. Either this man was, and is, the son of God: or else a madman or something worse. You can shut him up for a fool, you can spit at him, and kill him as a demon, or you can fall at his feet and call him Lord and God. But let us not allow any patronizing nonsense about his being a great human teacher. He has not left that open to us. He did not intend to.

But supposing God became a man, think our human nature, which can suffer and die, was amalgamated with God's nature in one person, then a person could help us. He could surrender His will, and experience death because He was man, and He could do it correctly because He was God. You and I can go through this process only if God does it in us, but God can do it only if He becomes a man. Our attempts at this dying will succeed only if we, man, share in God's dying. Just as our thinking can reach only because it is a drop out of the ocean of His intelligence: but we cannot share God's dying unless God dies, and He cannot be killed except by being a man. Is the sense in which He pays our debt and suffers for us, what He need not suffer at all?"

When a Russian cosmonaut returned from space and reported he had not found God, Lewis responded, "this was like Hamlet going into the attic of his castle looking for Shakespeare. If there is a God, he wouldn't be another object in the universe that could be put in a lab and analyzed with empirical methods. He would relate to us the way a playwright relates to the characters in his play. We, the characters, might be able to know quite a lot about the playwright, but only to the degree, the author chooses to put information about himself in the play. Therefore, in no case could we prove God's existence as if he were an object wholly within our universe. In like manner, those who claim biochemistry has proven man has no spiritual soul are equally misguided."

Lewis gives us another metaphor for knowing the truth about God when he writes, "I believe in God as I believe the sun has risen, not only because I see it, but because by it I see everything else. Imagine trying to look directly at the sun to learn about it. We can't do it. It will burn out our retinas, ruining our capacity to take it in. A far better way to learn about the existence, power, and quality of the sun is to look at the world it shows us, to recognize how it sustains everything we see and enables us to see it. Here we have a way forward. We should not try to look into the sun, as it were, demanding irrefutable proofs for God. Instead, we should look at what the sun shows us, which account of the world has the most explanatory power to make sense of what we see in the world and ourselves? We have a sense the world is not the way it ought to be. We have a sense we are flawed, and yet very great. We have longings for love and beauty; nothing in this world can fulfill. We have a deep need to know meaning and purpose, which worldview best accounts for these things?" [3]

A brand name is worth a lot of money. Before the advent of digital watches, every person who owned a real Rolex knew to keep it in his pocket when browsing a foreign flea market. A fake Rolex looks just like the real thing. A small-time shopkeeper will tell you his watches are genuine fakes, not one of those fake-fakes.

So, it is with Christianity. You better know your Bible because there are so many genuine fake and fake-fake feel-good Christian religions. Genuine Christians cannot be fooled by cults that pick and choose what they want you to believe. There is only one authentic Christian religion, just like there is only one authentic Jesus. My advice is to avoid splinter religions and feel-good spirituality that create their own brands to keep their flocks.

Jean-Paul Sartre (1905 – 1980)

Jean-Paul Sartre was a highly influential modern-day French atheist and existentialist philosopher who was awarded the 1964 Nobel laureate in literature. He believed everything about life was irrational and absurd. Living with emptiness is noble. Sartre rejected Descartes' proposition thinking gives meaning to existence. Instead, Sartre argued it is unreasonable to believe anything is rational. Sartre concluded life is so pointless, the only recourse is to act. Thinking and reason mean nothing. But by your actions, you at least know you are alive. It does not matter what you do as long as it makes you feel alive. You can feed a homeless man or bash his head in. It makes no difference. Why should you care? Moral reasoning is nonsense. There is no right or wrong.

Adopting words of Nietzsche, Sartre was "nauseated" by Christian morality, which robs a man of his freedom. In a bitter critique of Kant's argument on moral freedom, Sartre wrote, "man is condemned to be free." To Sartre, real freedom is indifferent to morality.

Sartre agreed with Kierkegaard's analysis of Abraham undergoing anguish, but claimed God told Abraham to do it; Sartre wondered whether Abraham ought to have doubted whether God spoke to him. In Kierkegaard's view, Abraham's certainty had its origin in "inner voice", which cannot be demonstrated or shown to another. "The problem comes as soon as Abraham wants to be understood." Why should we believe the voice in his head was God and not his guilty conscience? Sartre wants to free himself from such moral dilemmas.

Sartre demonstrated when it comes to philosophy; atheism is a hopeless undertaking. Without some form of metaphysics, atheism has no framework to explore why we are here, the source or meaning of existence. Atheism is its dead end. There is no reason whatsoever for us to be here. Sartre gutted existentialism of any substance. He is like an engineer who designs an automobile that requires no driver and accepts no passengers.

The problem is Sartre's atheism is a death pact with the devil. Sartre's atheism delivered a death blow to existentialism, for the end is not written. Our beliefs must rely on the fact of our existence and it takes no great step for us to assume we may all be a vital part of a broader phenomenon, searching for some more beautiful vision of life that could conceivably emerge from our present human condition.

Even if you disdain oxymoronic theologians as Sartre does, it must be conceded he had nothing better to offer. Sartre desperately wanted a better society, but the reality of flawed human nature so disillusioned him, he had nothing to contribute to making it better.

The existentialism of Kierkegaard opened a path to a better resolution of the contradiction's endemic to the human condition. Kierkegaard sought to reconcile God and man by coming to grips with the inconsistencies in common to both. Sartre rejected Kierkegaard's existentialism and the hope for a simple rational philosophy. Rather than furthering Kierkegaard's philosophy, the Christian intellectual movement turned to the liberal theology of Paul Tillich, which gutted the divinity of Christ. If there is to be real hope, if our real need is for a God with whom we can engage our lives, this would become a fatal turn of events to the traditional doctrine of Christianity.

Kurt Gödel (1906 – 1978)

His peers recognized Gödel as one of the greatest mathematicians who ever lived. He is most famous for the incompleteness theorem he developed in 1931. He proved there is no logic system that is self-contained, such that all possible propositions in the system can be determined. Gödel devised a thought experiment to prove his point. Consider a universal truth machine: When it is fed a true statement, it repeats the statement. If the statement is false, the machine is silent. Now feed the machine the following statement: "Two plus two equals five is false." The machine repeats this true statement. Now feed the machine: "I cannot say two plus two equals five." The machine repeats this statement. Now feed the machine: "I cannot say two plus two equals five twice. I cannot say two plus two equals five twice." This is a logical trap. If the truth machine says the statement, then the statement is false. If the machine remains silent, then the statement is true. Since the machine has contradicted itself, the machine cannot know everything about truth. Gödel concludes, "I know some truths the machine cannot address. Clearly, he has shown subtle ambiguities exist in meaning of language. But he goes further. If the universe is self-contained, it cannot contain within itself all knowledge about itself." Therefore, a more extensive system is needed, ad-infinitum. Rational thought can never know or explain all the truths of the cosmos. The cosmos can only be understood from outside itself. No finite being, no matter how intelligent, would have to know this. Therefore, throughout all space and time, no alien life can now exist or ever exist that would not know this. Gödel views the quest for knowledge as the universal goal of uniting all intelligent beings. God is the only universal truth that can be believed by finite people.

In 1941, Gödel [20] and Paul Oppenheimer [21] developed mathematical theorems which prove the necessity of God. These arguments are reminiscent of the ontological arguments of Anselm eight centuries earlier.

Murray Gell Man (1929 – present.)

Professor emeritus of theoretical physics at Caltech, Murray Gell Mann, is a most remarkable man. He received the 1969 Nobel laureate in physics for predicting the existence of the quark in 1965. In 1968, the quark was discovered at the Stanford linear accelerator. This began the search for a new generation of elementary particles, which are the constituents of protons and neutrons.

When he was growing up in Brooklyn, his parents were told he had a learning disability, and he needed to be held back in school. They were shocked because their son spent all his time reading library books about science. It turns out he was so bored with school he could not pay attention. They sent him straight to college, where he graduated from MIT with a Ph.D. in physics at age 22.

Gell Mann claims the existence of a well-ordered universe, in which sharp objects exist at distinct locations in space, and in which there is a well-defined concept of time, requires special cosmic initial conditions. His calculations indicate under nearly all initial states, a universe such as ours could not emerge. There would be no locality. In such a smeared-out world, one could know nothing without knowing everything. It is clear the source of creation being described is not some random explosion out of nothingness. If this were the case, Gell Mann and Stephen Hawking agree the universe would have been an amorphous blob of radiation, which would have annihilated itself very shortly after it emerged. The intelligibility of the universe, the fact we can progressively discover laws and extend our understanding of nature, the very fact science works would not be an individual and absolute right, but could be traced to and attributed to highly special cosmic initial conditions of the universe. Given these odds, any scientist readily acknowledges it is highly improbable our universe even exists. Stephen Hawking agreed. He wrote, "It would be very difficult to explain why the universe should have begun in just this way, except as an act of God who intended to create beings like us." A decade later, Hawking decided he had proof there was no God.

CHAPTER 5

COSMIC ORDER

"That's one small step for man and a giant leap for mankind!" Neil Armstrong, July 20, 1969. After landing on the moon, the astronauts read together the first chapter of Genesis.

Majesty of Physics

"It's difficult to avoid the impression a miracle confronts us here, or could the two miracles, the laws of nature and the human mind's capacity to divine them.—The miracle of the appropriateness of mathematics for the formulation of the laws of physics is a wonderful gift we need to understand what it is." Eugene Wigner, 1993 Nobel in physics.

"The most beautiful thing we can experience is the mysterious. It is the source of all true art and science." Einstein.

The greatest mystery of science is that it works! Scientists know we live in a rational, ordered cosmos subject to exacting harmonious and universal laws that can be understood by human reason. Why should inconsequential biological organisms have the intellect to discover and understand the nature of the cosmos from their own backyard? Is this a lucky fluke, or is it evidence of our inevitable connection to a higher cosmic power? It does not take rocket science or a leap of faith to recognize there is a profound relationship between the human mind and the cosmic order.

Is God big enough to stand up to the onslaught of scientific knowledge? Scientific knowledge is a God-given gift, guide, and passageway into the future. Science has demonstrated beyond a reasonable doubt the existence of man is uniquely essential to the nature of reality. With every discovery, science produces a picture of a more wondrous world. As it does, it appeals to our sense of humility. We are written into the laws of the cosmos in a profound and meaningful way. With all humility, we can stand tall, knowing we are meant to be here. Science does not and cannot deny there is meaning behind existence. It is up to each of us to open our hearts and minds to the discoveries of science as they pertain to the truth of our existence.

The persecution of Galileo will never be forgotten in the halls of science. It shows how hard it is to give up old thinking when challenged with new truths. Don't forget, Einstein couldn't accept the basic tenets of quantum physics. We must move forward for the sake of humanity, even if it takes generations dying off to correct past mistakes. It is clear any attempt to understand the nature of reality and the place of human beings, in reality, must now include a sound scientific foundation.

Reality is like an old puzzle from a yard sale. There is a picture on the cover, but is the same puzzle inside? How many pieces are missing? How many pieces are needed to confirm the same puzzle is in the box as on the cover? The Bible gives us the big picture, but it is just a guide to the puzzle in which we live. Science is putting pieces of the puzzle together. What makes the puzzle solvable is scientists can discover how pieces fit together one at a time. There is no logical explanation for why this is so, except God planned it that way for our benefit. The pieces could be so disconnected there would be no puzzle to solve, or the pieces could be so tightly interconnected the puzzle could only be solved as a whole.

Science is not the only method of thought that deserves our attention. Religion flourishes even in our age of science and technology.

The scientific quest is a journey into the unknown, which has never been revealed to mankind, except through Scripture. Believers must diligently seek truth and embrace science as a path to the truth of God's creation, without fear of what they may find. Each scientific advance presents unexpected discoveries, which challenge our minds with unfamiliar and complex ideas, but through it all, runs a solid thread of rationality and orderly progress.

Mathematics is a symbolic language, just like any other. Humans have invented thousands of languages, but there is only one universal language of mathematics because it arises out of nature. Neil deGrasse Tyson said, "Math is the language of the universe. So, the more equations you know, the more you can converse with the cosmos." Humans did not invent mathematics. There is no reason for math to be the language of the cosmos other than a great Mind like God thinks this way. Sir James Jean put it directly, "God is a mathematician." Pythagoras was the first person to declare God is a mathematician. It is fascinating math is the language of physics since it is an abstract language with no physical properties, except for the piece of paper on which it is written. Mathematician Richard Hamming, the inventor of computer error correcting codes, considers the discovery of integers to be one of the greatest achievements of humankind. He points out arithmetic is more useful in far more situations than a wheel.

The relationship between math and science is like something was prepared and left behind for us to discover by the Intelligent Designer. The laws of the cosmos are stunning equations, which are so elegant they can be written on a single piece of paper or a t-shirt. Math is the code that connects the mind of God to his creation. Mathematics is a gift from God that has opened up his secrets to science.

The intergalactic order was founded on beautifully conceived mathematical laws interweaved with each other to form a subtle and harmonious unity.

"Physicists abhor ugly theories and uglier math. The laws of physics are possessed of exquisite simplicity and have often commended themselves to scientists on the grounds of beauty and harmony alone. These same simple laws permit matter and energy to self-organize into an uncountable variety of complex objects, including those that have the quality of consciousness and can, in turn, reflect upon the very cosmic order that has produced them." [22] Paul Davies, Professor of Physics, Arizona State University.

Physics is the oldest and most fundamental science. All science ultimately reduces to physics. The best understanding of chemistry, biology, psychology, and neuroscience reduce to an understanding of the physics behind it all. As science explores ever deeper into reality, nothing is what it seems. The flat earth and moving sun are illusions. The earth's surface rotates 1,000 mph yet it seems like we are standing still. Even passenger jets flying at 300 mph cannot keep up with the sun, even though they appear to be whizzing by. The North Pole is no more up than the South Pole. Light from the sun looks white, but it is all the colors of the rainbow. We cannot see objects at all. All we can see is light reflected off of them. The object itself has no color at all. Color does not even exist outside the brain. Molecules don't have color, taste, or smell. Our minds make it up for us. Perception is not reality. The matter is just a bunch of point size atoms holding hands and a keep out sign. There is no sound without an ear to hear it. Feynman wrote, "A mite made the ocean roar." What our senses tell us is not there. Our minds create a virtual world for us. Even though reality seems the same for all of us, it can fool us when we need it the most. Lawyers know the mind makes a fool of all of us. A star that died millions of years ago can still be seen in the night sky.

Why do stars twinkle? Why does the moon have a backside? If Buzz Aldrin can jump up two feet on earth, how high can he jump on the moon? Why are the sun and the full moon always in opposite hemispheres? Why does the earth have a magnetic field? Why is the sky blue and sunsets red? What causes the Aurora Borealis? What causes rainbows? What makes a plane fly? Why is it so easy to balance a bicycle? How can cell phone towers handle thousands of calls at the same time? Why are no two snowflakes alike? Which is heavier, water or ice? Why does ice float? Why do clouds float? What causes lightning? How does water climb a tree? Would you like to know how computer chips, lasers, and holography work? The questions are endless and physics has the answers.

Cosmic Laws

"Everyone who is seriously involved in the pursuit of science becomes convinced a spirit is manifest in the laws of the universe, a spirit vastly superior to that of man, and one in the face of, which we with our modest powers must feel humbled." Einstein

"Who created those laws? Who made the laws of logic and mathematics?" "No matter how far the boundaries are pushed back, there will always be room both for religious faith and religious interpretation of the physical world. For myself, I feel much more comfortable in the concept of a God who is clever enough to devise the laws of physics that make the existence of our marvelous cosmos inevitable, than I do with the old fashion God who had to make it all laboriously piece by piece." James Trefil, physicist, "The Moment of Creation."

The laws of nature are the same today as they were yesterday and will be tomorrow. If we accept the laws of nature came from God, then one of the great mysteries of nature is explained. Otherwise, we must ask where they came from? Who wrote the instructions for an orderly creation process leading to conscious intelligent beings? The steady rise of the morning sun is a testament to God's faithfulness to life on earth. The assumption of inductive reasoning is the necessary prerequisite, which allows science even to be possible. As mere mortals, scientists realize they cannot know the true and perfect laws. Scientists know they exist because we have learned so much about them. They are the centerpiece of a scientific cosmic puzzle. They have proven themselves to fit together in a logical, harmonious and beautiful pattern, revealing nature's secrets, which have been hidden since time began. The parts could add up to a pile of incomprehensible nonsense. Instead, each piece of the puzzle has been discovered one at a time over the last few centuries, until they now form a coherent picture, which reveals the magnificence of the entire creation process. Max Born noted, "The continuity of our science has not been affected by all these turbulent happenings, as the older theories have always been included as limiting cases in the new ones."

Today, physicists have almost complete knowledge of the intimate connection between all the forces of nature and the rules which govern them. "There is a myriad of physical phenomena from the earth to the heavens, which form intricate relationships described by the laws of nature. Every discovery leads to more interesting and unexpected discoveries. The orderliness in nature revealed by science is there to be discovered. It is not at all natural that laws of nature exist, much less man can discover them.— This miracle may well be beyond human understanding." Schrödinger.

The laws of nature discovered in the last few centuries are imbued with the same attributes given to God thousands of years ago in the Bible. Science finds the laws of nature as they exist. There is no assumption of sacred properties in nature by science. It is just a marvelous and mysterious fact. Nothing is less obvious in the variety of and melee of nature than the existence of universal laws.

Newton's discoveries led scientists to realize the laws of the cosmos are the same laws of nature here on earth. There is no reason known to science why the laws of nature should be universally true. If there were no laws or they differed by time and place, nature would be incomprehensible. Nothing would make sense, but then again, we wouldn't be here to notice. There are no exclusions or immunities to the laws of nature. They are immutable, unchanging, and absolute. They have virtues which approach perfection beyond comprehension as they orchestrate the workings of the cosmos. The Grand Designer has given his laws to us in the precise universal language of mathematics. The fact the laws of nature are written in the language of mathematics has been first attributed to Galileo. "The miracle of the appropriateness of the language of mathematics for the formulation of the laws of physics is a wonderful gift, which we neither understand nor deserve." Merci Cooper, mathematician.

Absolute and immutable rule govern an ever-changing cosmos. The cosmos and every part and parcel within it can be described by its particular state. The state of any system or subsystem is defined by the specific values of its observable properties. As an example, consider a container of water. Theoretically, its state would be described by the particular state of each molecule. However, we are usually interested in its state as a whole, such as its temperature. Even more generally, we use its phase to describe its state. In this case, water is simply described as being in a solid, liquid, or gas phase state. In any case, a certain law of physics describes how the water changes its phase state with temperature and pressure. Every camping enthusiast knows water boils at lower temperatures at higher altitudes because atmospheric pressure is lower. At every moment in time, the state of the container of water changes according to conditions outside the container. This fundamental principle of science illustrates how physical laws that govern any physical system are separate and distinct from the system itself and the particular state of the system. The laws of physics describe how physical states change with time, but the laws themselves do not change with time or place. Absolute universal laws guide every physics event that takes place, but they never change. It should be clear why the laws of nature are omnipotent. They are universally all-powerful. Nothing in the cosmos escapes their power. Physicist Paul Davies goes even further, declaring the laws are omniscient because not only are the laws in command of all physical systems, "the systems do not have to inform the laws of their states for the laws to apply the right laws for every state of the system." [22]

Scientists now know the laws have an existence distinct and separate from the existence of the cosmos itself. The laws must have preceded the cosmos in order to cause the universe to be created. They existed before time began, and therefore transcend the cosmos. Science can only know the laws through their manifestation in the cosmos. Consider the analogy of a computer. The rules of the cosmos correspond to software, and the physical states of the cosmos correspond to computer hardware. We then want to know if there is independently existing software, which runs the universe. We see the software preceded the hardware. The existence of laws is as real as anything else we can know as real. The laws are more real than reality itself because they create and run the reality computer. Who is the Supreme Lawgiver of Genesis who would give us the privilege of knowing these laws? All known fundamental laws are mathematical formulations. This is a subtle topic that requires an investigation into the nature of mathematics.

Despite of the apparent chaos we call the cosmos, this chaos is nothing more than a façade hiding the many amazing symmetries inside. Symmetry is a fundamental principle of physics. Without it, the math of physics would be prohibitively more complex and lack elegance. For example, the description of the patterns of snowflakes is simplified by their symmetries. A sphere has perfect symmetry because it is invariant in shape, no matter how it is viewed. The human form has one degree of symmetry. The cosmos we perceive is like the surface of the earth, but as we dig deeper into the dirt, the rules of dirt become laws, and the laws become more ethereal, more symmetric, more harmonious, more elegant, more absolute, more perfect. "Nature is much more imaginative than we are." Lawrence Kraus, Professor of physics.

Author's note: Inspiration is a sure sign of the work of a great creative Mind.

Beauty Beyond Imagination

"Symmetries are deeply satisfying. They highlight an order and coherence in the workings of nature. The elegance of rich, complex, and universal phenomena emerging from a single set of universal laws is at least in part what scientists mean when they invoke the word beauty." [2] Brian Greene, Professor of theoretical physics.

It is nothing less than miraculous that the fantastic beauty of nature arose out of the apparent chaos of the big bang. The intellectual brilliance found in nature is of such magnitude it has challenged the greatest minds who ever lived. Today, science is the work of a network of global collaborators and competitors of the highest caliber. They are sincere, ethical, and dedicated professionals. The fantastic world in which we live is their inspiration and motivation. Scientists may not use the God word to describe their work, but they have plenty of excellent substitutes, from exquisite splendor to the majestic beauty of the highest order. Einstein stated, "The only physical theories, which we are willing to accept are the beautiful ones." We are talking about mathematical formulations from the simplest to the most advanced, which are inaccessible to all but a select few. But the general idea can be put into words, so here are some highlights.

The most amazing development in physics is the possibility of a theory of everything. This theory is the hot topic of physics research today. When completed, it will unify all the laws of the cosmos into a single coherent mathematical structure. This would be the culmination of a grand unification process, which began when Newton discovered the laws of motion and the law of gravity are universal and unchanging. Then Faraday unified electricity and magnetism. Maxwell completed this unification with the elegant theory, which recognizes light is an electromagnetic force field. Einstein unified space, time, EM, and gravity with his theory of general relativity. Stephen Weinberg [23] et al. unified electromagnetism with the electroweak force through the theory of quantum electrodynamics.

The elementary particles have been united into a standard model through the collaboration of many theoretical and experimental physicists. Complex experiments in giant particle colliders determine the 20 fundamental constants of the standard model of physics. Theoretical

GOD: The Evidence

physicists know these constants are not arbitrary. They are precisely tuned to make existence possible. Therefore, there must be an as yet unknown theory which predicts them. Another way of saying this is experience with the natural world leads physicists to expect majestic laws can only be attributed to a magnificent Mind. Quantum wave theory is in the process of being morphed into string theory, which is one of the leading contenders in the quest for a theory of everything. The goal is to unify quantum theory and general relativity into a theory of quantum gravity, which is the holy grail of physics. This theory must explain in great detail how we got from a non-existent cosmos at time zero to the universe we have today. It may very well be this theory is unique and inevitable, which has profound implications because it means the cosmos could not be otherwise. String theory is a mathematical framework, which has been studied since 1960, but so far has failed to be confirmed experimentally. A string is like a quantum wave, which can be likened to a vibrating violin string with the interior of the string hiding seven internal spatial dimensions.

"One cannot escape the feeling these mathematical formulas have an independent existence of their own, and they are wiser than even their discoverers, we get more out of them than we originally put into them." Heinrich Hertz, German physicist.

Cosmic Dawn

"God creates out of nothing. Wonderful, you say. Yes, to be sure, but He does what is still more wonderful: He makes saints out of sinners." Kierkegaard

"There is no known explanation of the special properties of our pocket cosmos." Leonard Susskind, Professor of Physics

Consider how physicists believe the cosmos began. [24] At time zero and before, the universe, as we know it, did not exist. There was no time, no space, no energy, no matter, nothing scientific observation could ever tell us. Instantly, the big bang burst forth into an unknown void.

99

History of the Universe

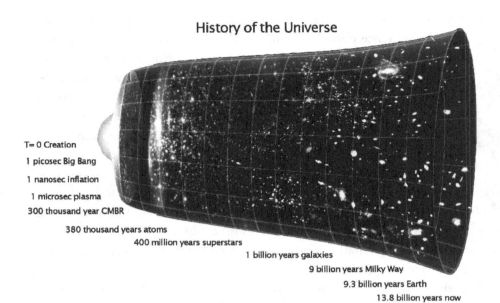

T= 0 Creation
1 picosec Big Bang
1 nanosec Inflation
1 microsec plasma
300 thousand year CMBR
380 thousand years atoms
400 million years superstars
1 billion years galaxies
9 billion years Milky Way
9.3 billion years Earth
13.8 billion years now

Today, the night sky reaches beyond our galaxy to illuminate the entire evolution of the cosmos, all the way back to its birth 13.8 billion years ago. Light travels through space unimpeded and undisturbed, preserving all the information it carries. At first, astrophysicists thought the big bang was a massive explosion of matter and energy. Over time, they believed the pull of gravity would slow down the expansion of the cosmos, so it would eventually collapse on itself. The big crunch theory was prevalent for decades. This theory postulated existence was an endless cycle of expanding and collapsing cosmos. This theory went out the window when it was discovered the rate of expansion of the universe was increasing rather than decreasing. [18] Saul Perlmutter received the 2011 Nobel in physics for this most unexpected discovery. The next theory falling by the wayside claimed universes were created in massive black holes. Then it was realized black holes are dead-end garbage dumps. Some theoretical physicists began to realize general relativity contains a vital clue that could explain the big bang.

The story begins with the familiar force of gravity. It is now known attractive gravitational force would have caused the early cosmos to collapse on itself. The cosmic background microwave radiation shows the old universe expanded in near-perfect equilibrium. A massive repulsive force was required to explain the big bang. Finally, it was realized general relativity predicted gravity would be repulsive rather than attractive under the conditions of the big bang. This effect is called antigravity. It gave rise to a new theory called inflationary cosmology, which was first proposed by physicist Alan Guth in 1981. His theory explains what happened at the instant of creation, time zero. In one Planck time interval, a Planck volume grew trillions of times, reaching trillions of degrees, culminating in the big bang. The enormous stretching and spreading created a uniformly distributed plasma soup of charged particles that would become the matter and energy to make stars. [25] More expansion occurred in this instant than in the rest of the time for which the cosmos has existed. The creation event itself, at time zero, is forever hidden by the plasma field of the big bang.

Professor Edward Tryon proposed the universe began as a quantum vacuum fluctuation where positive mass-energy is balanced by negative gravitational potential energy. The creation event is so small quantum theory must be applied simultaneously with Einstein's general relativity to explain the creation event, hence the theory of quantum gravity. According to the uncertainty principle and the de Broglie wavelength, space and time are not continuous; instead, they are granular! The smallest increment of time that can be located on the event horizon is the Planck time interval of 10^{-43} seconds. There are more Planck intervals in one second than there are seconds since the big bang. The smallest possible increment of space is the Planck length of 10^{-35} meters. This length is trillion times trillion times smaller than the diameter of an atom.

Creation began in a perfectly supersymmetric, zero energy state. At the end of the first Planck interval, the four fundamental forces of physics were unified into one supersymmetric super force. This is the scientific theory that the universe was created from nothing, i.e., creatio ex nihilo. The most insignificant nothing known by science is a perfect void. This is nothing in the sense there is nothing that can be considered physical or ever observed by science. But there must have been at least one supersymmetric quantum wave to get the ball rolling. This could not have been a quantum singularity of spacetime since the gravitational field would be physically unrealizable, i.e., infinite. The only possibility is this quantum wave released a massive barrage of quantum information. Quantum information is the universal code containing all the information necessary to create a baby cosmos.

At time zero, a whole supersymmetric super force spontaneously becomes unstable, and the symmetry is broken through phase transitions, which release the four cosmic force fields, beginning with antigravitational expansion. An analogy often used to explain supersymmetry is four people eating at a table. The situation is symmetric if each person drinks from the glass to his right. If one person drinks from the glass to his left, the symmetry breaks.

General relativity must apply during this interval to produce 4D physics. The big bang prepared the way for the synthesis of matter. Chen Ning Yang and Tsung-Dao Lee proved the concepts of broken symmetry for which they won the Nobel Prize in physics.

The lack of rotational forces in the early universe is one of the conditions favorable to the phase change from pure gravitational space to the big bang, a plasma of quarks, electrons, radiation, extraneous charged particles, and finally the simplest atoms. This theory also explains the cosmic asymmetry between matter and antimatter. This asymmetry favored the production of matter over antimatter by one part in a billion. This dangerous imbalance saved the cosmos from total collapse.

When the big bang plasma energy field cooled about a millionth of a second after the creation event, the collision energies had dropped to where the unstable elementary particles like quarks could begin to form protons and neutrons. At this time, all kinds of elementary and exotic particles, which are part of the present universe, existed, even though the temperature was still much too high for the formation of nuclei for the first few seconds. After the big bang, the curvature of space lessened the strength of gravity. It then became attractive and slowed the expansion down to the speed of light as atoms began to form.

The takeaway is the big bang was not a massive explosion at all. It was a perfectly orchestrated birth of a baby cosmos with all its excellent features in place. The big bang was like a rosebud opening up its petals to show its beauty. "The heavens declare the glory of God and the firmament shows his handiwork." Psalm 19:1

Einstein's cosmos

The creation story in Genesis was relegated to mythology for 2,500 years. In 1925, Einstein added a "cosmological constant" to his equation of general relativity to implement a steady state solution to accommodate an infinite cosmos. In 1929, Hubble published his work; which was scoffed at by astronomers for 50 years. When it was discovered Einstein's theory predicted the big bang if his cosmological constant were removed, Einstein said, "adding a constant to force his equation to predict time was infinite, was his biggest mistake." Now physicists are stuck trying to explain a finite cosmos. This is bad news for physics because infinity plagues mathematics and now infinity is out of bounds in physics forever. Nobel physicist Paul Dirac said, "The most important challenge in physics is to get rid of infinity." Feynman rejected theories that "brushed infinity under the rug." Physicist Frank Close states, "So finely balanced is the infinities in physics, extracting finite numbers by canceling them is like walking a tightrope over Niagara Falls."

Now cosmologists must think outside the cosmos. Their latest conjecture is the multiverse hypothesis. Theoretically, the number of universes in a multiverse must be finite because our universe is finite.

The proposed solution is like Rutherford's "plum pudding" model of the atom. The multiverse is an unbounded, amorphous void filled with universes. Or possibly the multiverse is like our cosmos with cosmos in place of stars. In this case, there would be a hierarchy of universes that go on absurdum, like Russian stacking dolls. Current simulations of the multiverse assume all universes obey the same laws, so variety takes place by tweaking 20 constants. Most universes created this way collapse from infant mortality. The laws of the cosmos are so well balanced; it is beyond comprehension to conceive of another set of rules that would create a universe with intelligent life.

If we take the Bible seriously, there are indeed other universes with totally different laws. One world with angels is highly recommended, while another is a hell hole of demons. With a little imagination, the laws of other worlds can be inferred by analyzing the incidents when intelligent beings from another world visited us as told in the Bible.

"Whoever has undergone the intense expanses in this domain of science is moved by profound reverence for the rationality made manifest in existence—the grandeur of reason incarnate in existence… Certain it is a conviction, akin to religious feeling, of the rationality or intelligibility of the world lies behind all scientific work of a higher order. This firm belief, a belief bound up with deep feeling, in a superior mind reveals itself in the world of experience, represents my conception of God." [26] Einstein

Back to our cosmic bubble, Einstein first envisioned a cosmos in a steady state of equilibrium. The cosmos did reach this condition 380,000 years after the big bang. At this time the universe had cooled down to the point the plasma of exotic, highly charged, unstable elementary particles, like quarks and leptons, i.e. electrons, could coalesce into neutrally charged stable atoms, a mixture of 91% hydrogen, 0.9% helium, and trace lithium gases. Suddenly, the plasma was gone, and the cosmos became a crystal clear vacuum. Gravity coalesced into massive gas clouds into superstars. 12 billion years ago, superstars began to burn out and die. In the throes of death, superstars collapse then explode into a supernova, which quickly burns out leaving large debris fields containing all the elements from hydrogen to uranium. Over a billion years, these fields coalesce into galaxies, i.e. our Milky Way. The fusion of two hydrogen atoms into a helium atom is the fuel that ignites a star into a fireball. Of all the particles produced by stars, it is incredibly fortuitous three helium atoms happen to fuse to form carbon 12, which is the most common element in stardust and happens to be essential to the formation of the incredibly intricate molecular structures that would lead to life. Hawking pointed out, if the electric charge of the electron was slightly different, stars would never have ignited.

Spaceship Earth

Our solar system was molded during the formation of the Milky Way galaxy 4.6 billion years ago. stellar dust converged by gravity into trillions of molten meteors and icy comets orbiting the sun. Their collisions formed eight planets, at least six dwarf planets like Pluto, 800 moons, the asteroid belt, the Khyber belt, and the Oort cloud. A small planet collided with the earth in a glancing blow. The earth's axis was tilted 23.5° off its orbital axis. The collision put the earth into high-speed rotation like a top. The remnant of the small planet went into orbit around the earth to become a moon. Because the moon is slightly oblong, earth's gravity captured the moon's rotation so only one side can be seen from earth. Most planets have elliptical orbits, but the moon's interaction with the planet puts the earth into a nearly circular orbit.

Thousands of comets pummeled the earth. As the earth cooled down, its surface was covered with an iron-rich green water world. Beneath the surface, molten lava cooled to a thick crust over the molten iron core. Thousands of volcanos belched reddish carbon dioxide, water vapor and toxic gases shrouded the earth in a dense fog.

By the time the earth was a billion years old, volcanic islands had reached the ocean's surface, and tide pools began teeming with stromatolites, a single cell organism like blue-green algae. For three billion years these cells multiplied by photosynthesis, a process, which converts toxic volcanic gas into oxygen while producing amino acids, the precursor of protein.

Oxygen oxidized the iron out of the oceans and purified the atmosphere. Nitrogen in the atmosphere made the sky blue. The earth became the blue planet. Lightning sends nitrogen to the earth where it fertilizes the ground. Multi-cell marine organisms emerged. The earth was on its way to becoming a very special spaceship, traveling through the cosmos at breakneck speed. The earth's surface near the equator rotates at 1,000 mph. The earth rotates around the sun at 70,000 mph. The sun revolves around the center of our galaxy at 450,000 mph. Our galaxy is moving away from the big bang at close to the speed of light. Quite a ride, considering you can sit in a lounge chair and feel none of it.

Scientists call our distance from the sun the habitable zone because the temperature range is just right for liquid water.

The size of the earth is big enough to hold its atmosphere, but not so big gravity would crush out life. The circular orbit has stabilized the environment, allowing time to start and sustain life. The tilt of the earth is just right to give us four seasons needed for growth all over the planet, top to bottom. An Aussie told me the North Pole is really a magnetic south pole, so we are the ones living upside down. The ozone layer captures heat from the sun and protects us from harmful cosmic rays. The earth's rotation gives us complex weather and ocean currents. The moon gives us tides, which are critical to marine life. Our massive neighbor, Jupiter, kindly protects us from catastrophic bombardment from meteors, comets, and asteroids. The rotating molten iron core forms a large magnetic field, which directs harmful radioactive and charged particles from sunspots and quasar mass ejections away from earth and sends them to the poles. Like rainbows and ice clouds, auroras are beautiful to see.

With an endless supply of recycled carbon, water, and minerals, evolution created a biosphere that went off the charts. Plate tectonics created mountainous continents with every kind of climate zone. Erosion from rain, rivers, ice, and snow shape mountains, green valleys, fertile plains, National Parks, Palm Springs, and Park City. Water tempers our climate and hydrates the landscape. Water in a gaseous vapor state is lighter than air, which is most fortuitous because it allows water to form clouds, which recycle water to land. If the solid-state of water were heavier than the liquid state, as with most liquids, ice would sink to the bottom of the oceans and lakes and the earth would become a permanently frozen snowball. The oceans are a massive heat reservoir, which delivers warm equatorial waters to the extreme latitudes by global conveyor currents, moderating weather over landmasses. The conductivity and PH balance of water is just right for life. Billions of years of living things buried by land upheavals have given us a massive energy resource of fossil fuel.

Mars once had oceans and atmosphere like earth, but was too small to hold on to them. But NASA recently found liquid water on Mars, which is the most important precursor to life. Meanwhile, let me tell you about earth's evil twin which is the same size and in the habitable zone. Venus started just like earth. But life never took hold. Over 10,000 volcanos shrouded Venus in greenhouse gasses. The surface temperature rose to 1,000° and water totally evaporated. Sulfuric acid and vaporized metals rain down, leaving shiny metallic covered mountains and rivers and lakes of acid. Russia dropped a satellite on Venus, which barely survived a couple of minutes. If billions of years of organisms on earth had not absorbed greenhouse gasses from volcanic activity, the earth would boil like Venus.

How could anyone not agree we are fortunate to be on the Goldilocks planet? The Kepler space telescope launched in 2012 has discovered 5% of stars in our galaxy have earth-size planets. As of this date, six planets have been found in the habitable zone, which are less than 2,000 light years away. Perhaps there are intelligent space aliens out there somewhere, but it is not a sure thing by any means. 6/20/21 Recent UFO sightings by the US Navy off the coast of San Diego caused a considerable stir because the pyramid-shaped objects were observed making maneuvers that defy the laws of physics as we know them, prompting the Director of National Intelligence to state, there is no evidence of any extraterrestrial activity. Plausible deniability has always been our government's policy on top-secret matters. With tongue in cheek, I would not be surprised to learn we may be experimenting with antigravity bubbles, as first demonstrated by Jesus. A habitable exoplanet must have an atmosphere, liquid water, magnetosphere, and stable orbit. But then again, there are about a trillion-trillion (i.e. trillion times trillion) stars and even more planets out there in a trillion observable galaxies. (Carl Sagan only knew of our galaxy.) There is even a solid diamond planet, or even perhaps a planet with rivers of gold. Newton loved alchemy, and he would have blown his mind. Oy vey! The Good Book says there is a city paved with gold out there somewhere. Who knew!

It is clear we would not be here unless we were written into the laws of the cosmos right from the beginning. How else could a planet happen to have ideal conditions for life?

As we enter the space age, it is helpful to know a bit more about how the speed of light affects space travel. On 2/18/21 the third mars lander traveled 300 million miles in seven months, Mars was 127 million miles from earth upon landing, resulting in an 11-minute delay in communication. The 11-minute time delay is referred to as the extended present; it is information happening concurrent to us, not in our past nor future. If we had a telescope powerful enough to see events on Mars, we would see the landing as it happens with 11 minute delay. This is how long it takes an event on Mars to enter our event horizon. At the time of this writing, Mars is 235 million miles from earth or 22 minutes from our event horizon.

The extended time or event horizon on earth is less than one-tenth of a second for every 2,000 miles, which is the same as human reaction time, so we think of telecom and TV as real time. Of course, there are time critical transitions. Preferred customers of wall street were allowed to have their computers in the same building. As a result, they could execute trades 100 times faster than a computer one mile away.

Miracle of Life

⫘⫘⫘

"DNA coded cells are so improbable as to appear to be a miracle." Francis Crick, 1962 Nobel Prize in Medicine

'The science versus religion debate is over." Dr. Craig Mello, 2006 Nobel Prize in Medicine

Biological evolution, is the most improbable progression in cosmic evolution. Even more improbable is the origin of life. Darwin clearly understood the creation of the first cell was not part of evolution. How several trillion atoms organized themselves into several thousand molecules forming the first living cell remains the greatest mystery in science. If life began spontaneously on earth, why has this event never repeated itself on earth sometime in the last four billion years? The discovery of DNA and RNA by Francis Crick and James Watson has given biologists the key to understanding how the cell works.

The fertilized ovum cell or zygote divides into 50 stem and progenitor cell types to form the embryo. The prototype cells construct the different body parts, organs, muscles, bones, skin, hair, a community of 500 to 100 trillion cells.

A living cell is a collection of proteins integrated into a complex unity of purpose. A cell processes more information than a supercomputer. Cells make decisions of all kinds. DNA is the master molecule of every cell. It contains all of the information passed onto each successive generation. The cell has three primary components: an energy source, a protective membrane, and a self-contained reproductive factory. Enzymes in the cell are responsible for screening sugars and breaking down glucose for energy. A replicator machine makes a copy of the DNA molecular structure, a transcription machine selectively segments DNA into tiny strands called chromosomes, which make up RNA, a splicing motor modifies and recombines RNA coded genetic information. A translation engine uses RNA coded instructions to synthesize proteins, and a molecular chaperone machine assists in manipulating and protecting budding proteins. Genetic mutations in humans occur about 60 times per generation. Most mutations are benign, but a few can be harmful or beneficial. A recent article in Science, co-authored by biologist Susan Lindquist, found mild hidden variations can become very active when a species must suddenly adapt when introduced to a stressful environment. Hsp90 (heat shock protein 90) causes genetic traits, which were once uniform, to suddenly show lots of variation. All mutations are passed onto the next generation. Every day, 50 billion of our cells die and are replaced, resulting in total cell replacement, except for neurons of the brain, every 30 years.

Palynology, anthropology, and genetics provide a modern and factual history of the evolution of the tree of life. All forms of life on earth share similar genetic codes and can be traced to a common ancestor; all living things are cousins. Even lowly plants communicate, cooperate, and wage war. When cattle were brought to Hawaii, many plants developed thorns to protect themselves. As far as we know, humans are the pinnacle of creation. Our brain is the most complex object known. The noble creature evolved from flawed, lower forms of life.

Single-cell organisms are the base of the trunk, and we are at the tip of a branch of the tree of life. The single-cell organisms used sunlight for energy in the process of photosynthesis. These cells combined to make multicellular plants from seaweed to grass to trees. Then cells evolved from invertebrate sea organisms to vertebrate fish, which use oxygen for energy. Fish moved onto land, evolving into reptiles, which evolved into mammals, like our primate ancestors. Humans and apes share 97% of the same DNA. All human DNA is the same within 99.7%. Every continent has a distinct diversity of life, whereas modern humans are a unique global species.

There are 37 million living species; thousands of new species are discovered every year; a billion species are extinct. Evolution is a cooperative endeavor of symbiosis.

No one knows how mindless molecules create the will to live as explained by the survival instinct. Biologist Richard Dawkins wrote, "The living results of natural selection overwhelmingly impress us with the appearance of design as if by a master." The living cell certainly has all the elements of intelligent design. Hawking stated, "The very purpose of science is to discover design in nature. The secrets of life are the source of knowledge beneficial to humanity. Mindless atoms are a critical component of universal design, but they do not deserve credit for their design, let alone the design of living things. Indeed, design in biological systems is subtle, but has proven to be a prolific source of biodiversity essential to our survival.

Let's take a look at how evolution accomplished this fantastic feat before deciding it is nothing more than a random accident. Biological evolution is a goal-driven search strategy, just like the evolution of human knowledge. Consider the social process of gaining experience and knowledge. One out of 1,000 people has a new idea. All but one of these ideas fall by the wayside, like the Ford Edsel. Only one out of 1,000 ideas (i.e. analogous to a good mutation) is successful, like the Ford Mustang. Each generation builds on the successes of previous generations, which contributes to the success of the next generation and a better future for the human species. The knowledge gained from successful ideas in the last 50 years has surpassed the rest of all prior experience. Evolution is the mechanism of creativity, invention, and discovery. So, it is with biological evolution. Nature's highest achievement and the end goal of humanity proceeded along an evolutionary path, just like all of life. The real insight is everything we know comes from understanding nature. Nature is the real genius, as pointed out by Einstein many times over.

In the world of computers, artificial intelligence uses step-by-step trial-and-error strategies to find approximate solutions to the broad class of problems, which are too large or too complex to be solved by any known deterministic, or optimal algorithm, or closed form computational method. The problem can be thought of as traversing a tree starting at its root. The tips of each branch represent all possible solutions to the problem. The AI search algorithm searches every branch of the tree, looking for a path along a branch whose tip has the best solution. A directed search uses pieces of problem-solving knowledge to prune branches from the tree and thereby reduce the size of the search tree. The tree of life is like a search-tree with all existing creatures on the tip of a branch.

In chess, each move a player makes is along one of the branches of the search-tree, consisting of all possible moves. In 1996, IBM's chess-playing computer, Deep Blue, defeated Garry Kasparov to become world champion. The search algorithm could look eight moves ahead, while even the best players can see about 3.

As a simple example of a directed search, consider how a GPS navigation system finds the best route through a grid of city streets. You enter where you want to go, i.e., goal state. At each possible intersection or decision point along the route, the algorithm tries each possible next step along the road to find which step on the route gets you closest to the destination. In this way, the best route is found incrementally or step-by-step.

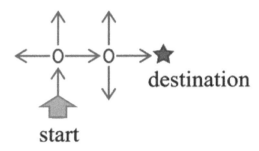

Steven Weinberg used the following example to explain how scientific research proceeds. "A party of mountain climbers may argue over the best path to the peak, and these arguments may be conditioned by the history and social structure of the expedition, but in the end either they find a good path to the peak or they do not, and when they get there, they know it." [23] Like the search-tree, the hill climbing analogy is fundamental to all learning strategies.

An example applicable to biological evolution has multiple possible hill tops or goal states, as depicted below. This particular problem-solving strategy is analogous to the evolutionary process in nature. The horizontal base represents an environmental search space starting with single-cell organisms at the bottom. The hilltops represent optimal survival states for five distinct possible species. All possible species begin in the primordial soup at the bottom. Each successive generation is dominated by those organisms which move up the survivability hill. When an organism reaches a survival goal state for a particular environment or hilltop, it becomes a stable species fully adapted to its environment. Species C achieves a goal state through the most adaptation, while species A and E needed to adopt the least to survive in their environmental habitats. This hill climbing metaphor is precisely analogous to the search-tree problem. In this case, every branch tip represents a viable species. Known as the Tree of Life, it is built using DNA to follow the development of every species on the tree, including humans, the king of the mountain on earth. Of course, this knowledge was unavailable to Darwin, but it is the smoking gun that conclusively proves biological evolution is a brilliant design strategy.

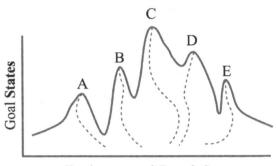

The number of problems which can be solved by this method is limitless. Evolution is the most robust learning strategy known to humans because it gains knowledge incrementally as it seeks a problem solution, rather than requiring the entire solution to the problem to be known beforehand. The entire universe evolved from nothing. If the universe is not the result of intelligent design, the only alternative would be a stupid design or no design, in which case there would be nothing for scientists to do but twiddle their thumbs.

Science does not yet know how life began. All we can say is it appears to have happened like everything else, through the natural creation events according to God's foreordained plan. The grandeur of God's works commands awe and reverence regardless of what processes created it all. Believers should embrace science, especially evolution, for what it has done to reveal the magnificence of divinity in-depth never dreamed by our ancestors. We have learned a lot since the advent of science and knowledge should never be dreaded or denied. Instead, science should be welcomed by all who cherish human knowledge and understanding. Skeptics like Dawkins go to great lengths to prove there are hidden flaws in the design of the human body as proof we are a product of evolution, not intelligent design. While this is true, it only serves to illustrate God has a higher purpose for us than mere creature perfection. The following passage makes it clear God created all life, including man, from inanimate particles of dust.

"Streams came up from the earth and watered the whole surface of the ground, the Lord God formed the man from the dust of the ground and breathed into his nostrils the breath of life and man became a living being." Genesis 2:6, 7

This scripture foretells the notion that life was conceived out of the primordial soup. However, the story also clearly conveys the special nature of humans who were created imago Dei, meaning humans have an eternal soul, about which Darwin knew nothing. The first creation story relates how humans were created on the sixth day as the ultimate purpose of creation. God does not create people through a simple command like the rest of the universe. He does not say, "Let there be man." He precedes the creation of humanity with the unique phrase, "Let us make man in our image." Evolution does not explain how life began, but it does explain the astounding diversity of life forms in every niche environment on our planet. "Thank God for Evolution." [27] Reverend Michael Dowd, Christian Evangelical Minister. "Saving Darwin: How to be a Christian and Believe in Evolution" [28] Karl W. Gilbertson

Extraordinary Design

"My religion consists of a humble admiration of the illimitable superior spirit who reveals himself in the slight details we can perceive with our frail and feeble mind." [16] Einstein

The magnificent design in nature is nothing less than awe-inspiring to those who take the time to notice. Every field of science has benefited humanity by discovering the secrets hidden in nature. The incredible discoveries of science should serve to give us a greater appreciation for the wonders of nature than our ancestors had. We have barely left the confines of our planet, yet we already know the secrets of the smallest particles and the farthest galaxies.

Design, whether by humans or in nature, is the purposeful arrangement of parts. Such planning is a requirement of the design. We infer design whenever elements appear arranged to accomplish a given purpose. According to physicist Hugh Ross, "The complexity of the cosmos is 10^{120} times greater than the complexity of the most sophisticated technical accomplishments of humans." Every discovery of science is leading us down a strange path, which is becoming so inaccessible and mysterious; it is reasonable to conclude it is leading us to God's doorstep.

Madame Marie Curie received the 1903 Nobel in physics and the 1911 Nobel in chemistry. She was the first woman to receive a Nobel in science and the only person to win twice. She said, "A scientist is a child confronting natural phenomena that impress him as though they were fairy tales." In so many ways, the fairy tale mysteries in nature reflect the fairy tale mysteries in the Bible.

Let's revisit scientific evidence that supports the biblical view. First, there can never be evidence that falsifies the God Hypothesis. Second, there is no evidence supporting the only possible competing hypothesis, mindless accidents. The laws of the universe represent such an astonishing number of fortuitous coincidences that the senseless accident hypothesis requires a lot of hand waving. The simple fact is the probability we are here by cosmic accidents is so infinitesimally small; it is not only statistically insignificant; it is virtually zero. Atheism has no foundation in scientific fact. Atheism is the false certainty life has no real meaning. True religions are open-ended faith that we are here for a reason, and our lives have meaning and purpose.

The more deeply science has delved, the more astounding the existence of our Goldilocks planet becomes. It is no longer enough to say we shouldn't be here. From what we now know, even our universe should not be here. And beyond our world, there is no plausible scientific explanation for why anything exists, period!

The emergence of the universe from its inauspicious beginning into a self-organizing, self-sustaining progression of increasingly complex entities should be recognized for what it is, a genuine extraordinary, design. Theoretical physics has inspired the development of ever more sophisticated mathematical methods to represent the most intricate details and grandiose features of nature. Understanding physical processes in nature involves every form of mathematics to capture the symmetries, diversities, and complexities from a spiral seashell to a spiral galaxy. Einstein wrote, "How can it be that it is possible that mathematics, a product of human thought, which is independent of experience, is so admirably appropriate to the objects of reality?" He knew science has brought us full circle. The creation story in Genesis and the metaphysics of Plato belong on the library shelves next to science books. The Creator spoke the universe into existence in the language of mathematics. He is letting us learn his language, the language of the tree of knowledge. The knowledge of physics is not just the language of creation; it is the cosmos itself. The cosmos exists only as a creation within God's Mind. The greatest mystery of existence has been laid bare. The mysterious God of the Bible has given us his final revelation of himself. The truth of physics, metaphysics, and the Bible have merged into a coherent picture explaining how we got here and why we are here. Every religion in the world has revealed some truth about God, from Eastern mysticism to Christianity. Every religion in the world has touched the elephant of the cosmos.

Our universe evolved like it was a box of scrambled puzzle pieces that self-organized until it emerged into a beautiful and harmonious picture by both logical and fortuitous selection of pieces. This cannot be characterized as pure luck. It is the necessary precondition of the cosmos to allow for our arrival. Scientists have uncovered the transcendent nature of the laws of the universe. They are omniscient in their power to control every aspect of every detail in every corner of the world. The ethereal nature of existence itself is a baffling mystery to science. Our universe is so unique the form it has taken is not the biggest mystery after all. The fact the world exists at all is the most baffling and difficult challenge to people of reason. The fact we have a mind capable of comprehending this fact is equally baffling and inexplicable. So, it is not surprising God is baffling and incomprehensible too.

All scientists readily agree nature shares our appreciation for economy, efficiency, harmony, beauty, logic, mathematics and subtle complexity. Words like ingenious and clever are undeniably human qualities, but it is undeniable they exist in nature too. It cannot be explained except as genuinely intrinsic to nature. Einstein called it a "cosmic religious feeling." We are creatures born of stardust, yet we can ponder our place among the stars. We are intimately related to the totality of the cosmos.

Biologists now know how to use evolution to design life, just like we are little gods made in the image of the eternal God. We should be careful about what we wish because we have been given immense power. It is better to remind ourselves that without God; we genuinely are worthless chemical scum.

"I cannot believe our existence in this universe is a mere quirk of fate, an accident of history, an incidental blip in the great cosmic drama. Our involvement is too intimate. The physical species homo may count for nothing, but the existence of mind in some organism on some planet in the universe is surely a fact of fundamental significance. Through conscious beings, the universe has generated self-awareness. This can be no trivial detail, no minor byproduct of mindless, purposeless forces. We are truly meant to be here." Paul Davies, Professor of physics and mathematics.

"How marvelous and intricate life turns out to be! How deeply satisfying is the digital elegance of DNA? How aesthetically appealing and artistically sublime are the components of living things, from the ribosomes translates RNA into protein, to the metamorphous of the caterpillar into a butterfly, to the fabulous plumage of the peacock attracting his mate. Evolution, as a mechanism, can be and must be true. But it says nothing about nature or its author. For those who believe in God, there are reasons now to be more in awe, not less." [29] Francis Collins, Director of the National Human Genome Research Institute, Director of the National Institute of Health, National Medal of Science.

CHAPTER 6

QUANTUM WORLD

"I don't like it, and I'm sorry I ever had anything to do with it."
Schrödinger

Foundation of the cosmos

"The cosmos is weird, and the earth is freaky. We shouldn't be here."
[25] Hans Pagels, Physics professor

A renowned architect and atheist constructed an imposing monument to modern science to banish religion. The building featured columns supporting nothing, a staircase to nowhere, windows without walls, and walls without windows. A curious rabbi stopped to look at the unusual project. The architect explained to him the structure symbolized a cosmos without purpose. The rabbi thought about it a moment, then replied, "What about the foundation of the building? Does it not have a purpose?"

Richard Feynman was born in 1918. He received the Nobel laureate in physics in 1965. His lectures on physics at Caltech are the stuff of legend. When I was a student there, hardly a day went by his name was not heard. He is an icon at Caltech. In the evening, his favorite pastime was visiting student housing to entertain with his knowledge, wit, and bongo. He was proud to be an atheist, like his father. Most people will remember the flamboyant scientist testifying before the Congressional hearings on the space shuttle Challenger disaster on 1/28/1986. Professor Feynman waited patiently for his turn, as senators and generals played to the TV cameras. Finally, it was his turn. Placing a rubber band in a glass of ice water, he ingeniously demonstrated the O-rings of the shuttle's fuel tanks had failed because rubber is brittle under icy conditions. The hearing droned on, but Feynman had achieved national celebrity. A year later, Feynman was fighting cancer at age 69. By chance, he ran into an old acquaintance. Herman Wouk was a devout Orthodox Jew from Minsk, Russia, the place of Feynman's ancestry. An acclaimed writer, Henry received the Pulitzer Prize for his novel "Caine Mutiny," and he recently published "The Language God Talks." [30] The title recalls his first encounter with Feynman 30 years earlier when Feynman whimsically said, "Calculus is the language God talks." Stoically facing the fate before him, Feynman listened while Wouk recalled for him his first TV interview:

"It does not seem to me that this fantastically marvelous cosmos, this tremendous range of time and space, the different kinds of animals and all the different planets and all these atoms with all their motions and so on, all these complicated things could merely be a stage so God can watch human beings struggle for good and evil, which is the view religion has. The stage is too big for the drama."

Wouk: "Do you still believe it?"

Feynman: "I see no need to change it."

Wouk: "You ad-libbed the power of Psalm 104."

Feynman: "You are putting me on."

Wouk: "You ad-libbed the power of wonder in a child, the wonder of philosophy, the wonder of science, of religion. Religion lives within the hearts of men."

Feynman: "Well, I enjoyed myself."

The Christian reply to Feynman would be only the life of Jesus provides drama fitting to the size of the cosmic stage. Like most people nearing the end of life, Feynman turned his thoughts to questions with ultimate meaning. Ages before science began, religion had claimed the high ground, but Feynman had never been interested in old ideas. The ancient battles of the gods had been ceded to the Jews. Their brilliant minds had struggled through bedrock issues to find one God big enough to satisfy minds like Feynman's. But he could not accept the paradox of an all-powerful Creator caring about flawed creatures on an insignificant planet.

Feynman had rejected religion when he was ten years old. He harbored a lifetime of disdain for feeble religious minds, with no real knowledge like science. His true love was quantum physics, the rock bottom of science. Quantum physics is a puzzle wrapped in an enigma. Feynman once said, "Quantum mechanics describes nature as absurd from the point of view of common sense."

Science is a study of what we can perceive in the physical world with our five senses. Scientific instruments are just an extension of our perception, such as telescopes and microscopes. Science has now seen the beauty of nature at every scale, from the far reaches of the cosmos to the inside of an atom. By the beginning of the twentieth century, physicists had discovered the elegant simplicity of the atom. But quantum physics soon created havoc. Suddenly, nothing made sense. The bedrock of the cosmos was sitting on nothing but ethereal quantum waves.

Physicists know they are staring at a whole new version of reality, which was the antithesis of everything they knew from classical physics. For the first time, physicists know they are not just doing physics, they are struggling with the very meaning of reality. The field of theoretical physics is finding itself asking questions formerly reserved for philosophy and religion. Physicists now want to know what it means to exist. The question of how existence began is now clearly understood as a why question. If a tree falls in the forest and no one hears it, did it make a sound? Yes! An insect will hear it. If no one is looking at the moon, is it still visible? If a big bang happens but no one knows it, did it still happen? Feynman's mentor at Princeton, John Wheeler, famously said, "The cosmos requires an observer to exist." If the cosmos existed for billions of years without conscious beings, who was there to observe it? Wheeler believes in the omniscient Observer.

In 1954, physicists set out on a new mission, to find out how protons and neutrons are made. Huge particle accelerator-colliders smashed these particles into more elementary particles with exotic names like quarks, leptons, and bosons. By 1966, it was realized a key member of the family was missing, dubbed the "God particle," and even bigger colliders were needed to find it. Also, a theory was needed to explain this deeper level of physical reality. In 1980, this new theory was tentatively crowned the theory of everything. It would bring general relativity and quantum physics into a unified theory of quantum gravity. String theory offered an elegant way to predict the particles of the standard model known only through experiments. The theory predicts there are seven hidden dimensions within a string. Professor Leonard Susskind calls unseen dimensions the "DNA of the universe." Meanwhile, another promising theory using a different approach has emerged, called loop quantum gravity. A critical missing particle, the long sought-after God particle, was discovered in 2012 at the CERN laboratory in Switzerland. Physicists are becoming more inclined to contemplate the deeper meaning of their work as they explore the mysterious foundation of the cosmos. To get to the bottom of existence, physicists are being forced to examine the very nature of reality. "Reality is not what it seems." [7] Physicist Carlo Rovelli Theoretical physicists have known for a century reality is just an illusion, however persistent and real it seems. Let me give you an idea of what this is about. The atoms, which form matter are energy force fields, which react with space and time through the forces of gravity and electromagnetism (EM). The EM field of light gives visibility to matter. Newton's second law tells us, "For every action, there is an equal and opposite reaction." Kick a rock, and its force field will kick you back with the same force. Ouch! The rock is innocent. You caused this to happen. Does the fact it hits you back mean there is something real there? Sometimes, things you can't see hit you back, like glass for instance. Just because you feel something only means your mind is responding to a force. There is nothing else there. When you see something, you don't see the thing itself, you only see the light reflected off it. Some of the light is converted to heat in the object. Shine a red light on a blue object, and it will be invisible. Objects do not even have a physical property called color. Color is just the frequency property of light, which we perceive as color. The same can be said of taste and smell. 95% of the matter and energy in the cosmos are invisible to astronomers.

Niels Bohr insisted there is no quantum reality beyond what is revealed by an experiment, an act of observation. Physicists now realize the reality we perceive is only an illusion of our limited perception. Physicists now know the foundation of the cosmos is the information of the universal cosmic code, which has all the instructions necessary to create what we perceive as reality. The underlying nature of physical reality is an ethereal foundation with no physical properties that can be observed in any experiment now or in the future. It is hard to escape the conclusion science has arrived at God's doorstep. God spoke the cosmos into existence. Genesis states God's words are the foundation of the universe. God said, "Let there be light and there was light." Genesis 1:3 This is only one astonishing fact in Genesis, which has been confirmed by science. There are many more to come.

God Particle

"The Higgs boson is the last missing piece of our current understanding of the most fundamental nature of the universe," Martin Archer, physicist

View in 360° Detector Assembly

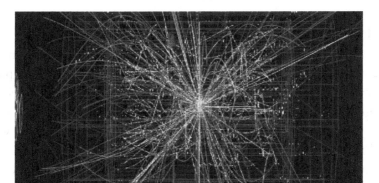

The cosmos can be described as a layered hierarchy of articulated entities from the largest to the smallest: galaxies, solar systems, living cells, molecules, atoms, subatomic particles, elementary particles at the bottom. Cells are the building blocks of life. Molecules are the building blocks of matter. Atoms are the building blocks of molecules. Protons, neutrons, and electrons are the subatomic building blocks of atoms. The elementary particles are the building blocks of subatomic particles.

The standard model of elementary particle physics is composed of six massless messenger particles called gauge bosons and six families of particles with mass called quarks and leptons. Protons and neutrons are called hadrons. They are made from three quarks and a gluon boson which holds them together.

Scottish physicist Peter Higgs predicted the Higgs boson in 1964. The Director of Fermi Lab at U. Of Chicago, Leon Lederman, led the search for the Higgs boson for 30 years without success. He called the Higgs the "Goddamn" particle, but his publicist convinced him to call it the God particle. The high-energy particle collider at Fermi Lab was not powerful enough to produce the elusive Higgs. Congress funded a bigger collider in Texas but defunded the project while the foundation was being laid.

Instead, the EU funded the European Council for Nuclear Research, CERN, located near Geneva, Switzerland. It is the workplace of 10,000 scientists and engineers representing 100 countries and 600 universities. The $10 billion Large Hadron Collider, LHC, at CERN is the most powerful collider in the world, at 7 Teva (Trillion electron volts). It is a ring of 1,600 thirty-ton magnets with more iron than two Eiffel towers. It is 328 feet underground in a 17-mile circular tunnel. Each magnet is cooled to 4° Kelvin. Helium filled pipes produce superconductivity, which dissipates heat to allow extreme levels of high electrical currents to energize and control the magnets. The LHC is fed by two reaccelerates, a proton synchrotron, and a super proton synchrotron. The particles produced in the high energy collisions are analyzed by way of a dozen separate pathways to 360° detector assemblies. To produce the elusive Higgs boson, which has a lifespan of less than a trillionth of a second. A proton and antiproton traveling within 0.0000001% of the speed of light in opposite directions collide billions of times per second to observe a few Higgs bosons. The colliding protons are broken into quarks; then among random quark collisions, a direct hit is required to produce the Higgs. Only one out of a million proton collisions result in producing the Higgs because it needs a perfect head-on collision between two quarks. The Higgs must then be sorted from all the other exotic particles ejected by the collisions. The discovery of the Higgs boson was announced with much fanfare on July 4, 2012. Peter Higgs received the 2013 Nobel in physics.

So now we owe our very existence to the reverse process, the production of protons from the pure energy fusion of Higgs bosons into quarks. The Higgs boson is the quantum of the Higgs field (aka Higgs ocean). The Higgs field is a non-zero vacuum energy state filling all spacetime, providing a source of apparent mass-energy for the other elementary particles. The existence of the Higgs field results in spontaneous breaking of the electroweak gauge symmetries. Because the Higgs boson has zero spin it is susceptible to the other frenzied vacuum fields around it. The Higgs field is invisible to massless particles, such as photons and gravitons, but it is like a viscous fluid to other particles, like quarks and leptons. The Higgs gives these other particles the property of apparent mass. Now the Higgs boson has been found, the specific processes of mass production by the Higgs field can be studied. The Higgs breakthrough is expected to eliminate the concept of mass itself as an inherent property of matter.

5/2013 TED Conference @ CERN: Theoretical physicist Gian Giudecca stated, "The Higgs field may be standing on a knife's edge, ready to collapse. This is a hint our universe may be only a grain of sand on a beach of multiverse." He went on to say the mass-energy of the Higgs boson has been measured to be 126 GeV (billion electron volts). This value is precisely particular because theoretical models predict the Higgs field is balanced between two massively different phase states. It is predicted a sudden quantum transition between these two states would initiate an avalanche of self-destruction of our universe. Fortunately, the odds of this happening are about one in 10^{150}.

The Higgs field was expected to explain why particles come in very different sizes. Recent results indicate this is not happening. Physicists are now convinced there is a deeper mystery yet to unfold. Since 2015, the energy of the LHC is being doubled to 14 Tev, to search for even higher energy particles. What physicists don't know is what they want to know. The usual course of physics before the twentieth century required a theory to follow the experiment. Maxwell and Einstein changed that. Also, the cost of experimental physics is becoming a significant hurdle to progress. Experimental physicists are becoming impatient because theoretical physicists are leaving them behind. There are just too many unanswered questions. Why is there stuff? Why are there all these strange particles with precisely the right properties to make stuff? Why do we have these particular physical laws, or even something called "physical law" at all?

In a lecture titled "Is the End in Sight for Theoretical Physics?" Hawking said, "There are grounds for cautious optimism that we may now be near the end of the search for the ultimate laws of nature." As far as theories go, string theory has been the leading contender for the theory of quantum gravity for over 50 years. It may be another decade before there is experimental confirmation. String theory is a natural extension of quantum physics.

Experimentalists are having a hard time giving up on the idea that the fundamental particles are not particles in the sense they do not occupy a point in space. This leads us into the topic of quantum field theory, where the very definition is not a particle but a quantum wave.

Theory of Everything

"My equation is smarter than me. God is a mathematician of a very high order, and He used advanced mathematics in constructing the universe." Paul Dirac, 1933 Nobel in physics

The search for a theory that would unify the four forces of nature has been at the forefront of physics for over a century. The specific challenge has been to find a theory that would bring general relativity into the world of quantum mechanics. The first breakthrough took place in 1928 when Paul Dirac published relativistic quantum field theory. NASA's Fermi Gamma Ray Space Telescope has discovered electrical fields in space, which provide experimental confirmation of Dirac's theoretical work. Now it is known the vacuum of space only appears to be empty. The vacuum is boiling into spontaneous eruptions of electrons and positrons, which vanish as quickly as they appear. The constant interaction of electrons with positrons in space is a cause of familiar static you hear on your radio and TV when they are not tuned to a signal from earth. The fluctuating fields of vacuum space are called the Dirac field, or euphuistically, the Dirac sea. The Dirac sea is the source of all the electrons in the cosmos.

Dirac's theory proves the vacuum of space is filled with an undetectable field of virtual, i.e., potential electrons. Energy perturbations in space can cause virtual electrons to become real ones. Each time this happens, a positive hole is left behind in space like a ghost. These holes behave like real particles, called positrons, which are the negative matter with the opposite electrical charge of the electron, moving through the vacuum of space. Electron-positron pair annihilation occurs when an electron falls back into a hole in the Dirac field. It is important to note a quantum particle cannot have zero energy. Electrons released in free space are in their ground state, which is the vacuum energy of space. Energy is measured by temperature in physics. The vacuum energy of space is 3° K.

Now there is a prototype for relating general relativity to quantum mechanics. The negative energy of gravity releases quantum particles with positive energy and a negative charge called electrons. Electrons are the quantum particles, which create or absorb the photons of electromagnetic fields. Of the four cosmic forces, three have now been combined into a grand unification theory, GUT. The remaining challenge is called quantum gravity. When integrated into the GUT mathematical framework, the theory of everything TOE will have the formal name of relativistic quantum field theory or simply quantum gravity.

Since 1975, string theory has been considered the most promising path to the holy grail of physics. While the general principles of quantum gravity are well understood, thousands of research papers have come up empty-handed. Now experimental results at CERN have found a black swan. String theory predicts a whole new set of complementary symmetric particles, but none have yet to be found.

To summarize where we are, from general relativity, it is known the negative energy of gravity shapes the fabric of spacetime and counterbalances the positive energy of matter, such that the cosmos began from zero energy and the sum total of all energy is still zero. [25] Conservation of energy means our cosmos is isolated from its exterior, no energy enters or exits. The graviton, as yet unobserved experimentally, is the quantum of gravity. Gravitons are massless and chargeless messenger particles, which transmit changes in the gravitational field throughout the fabric of space at the speed of light, just as photons transmit EM fields. The four cosmic forces create four quantum fields: gravity, EM, Higgs ocean, and Dirac sea. Technically, gravity is the spacetime geometric container of the cosmos, an elastic web which simply reacts to the other quantum fields within its boundaries. The corresponding quantum particles are the graviton, photon, Higgs boson, and the electron. The graviton and the photon carry information but have no mass. The Higgs boson creates quarks, which combine in triplets to make protons and neutrons. Protons and electrons have four properties: mass, charge, spin, and momentum.

A more recent approach to TOE, the Wheeler-DeWitt equation, called loop quantum gravity, is getting all the attention today. Like string theory, loop quantum gravity applies quantum theory to describe what happens when the equation of general relativity reaches a singularity at the creation event and black holes. Loop theory postulates gravity is a fabric made of Planck size loops, hence the name loop quantum gravity. The theory predicts the first step in the creation of a galaxy is the collapse of a dying megastar into a Planck size star. A Planck star is a billion to trillion times smaller than an atom. From this point, antigravity explodes into a supernova, which quickly collapses by normal gravity into a massive galaxy of smaller stars with a black hole in the center.

The creation event is entirely different, but it does start at the size of a Planck volume. The difference here is gravity could not have existed before creation because the results are entirely different from a collapsing star. The antigravity event at conception had perfect symmetry and zero entropy, which are quite the opposite in a plank star and black hole. Theoretically, our universe could have been created by a collapsing world, but this seems highly unlikely because our universe shows no signs it will ever collapse.

Loop quantum gravity postulates gravity is an elastic matrix like the web of various grains of Plank-size volumes shaped like tetrahedrons, which allows it to stretch, shrink, bend, sway, break and fold as gravitons interact with the particles of the other quantum fields, which are bound to the fabric of gravity. The graviton has the property of spin and gravity is described as a spin network.

3D Tetrahedron

 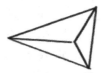

The nodes of the field are gravitons, which are linked tightly by a 3D grid or matrix of gravitational field lines. These Planck size loops differ from magnetic field lines, which are parallel and never touch each other. The Planck volume is the unit of spacetime, which is a property of gravity. The general relativity of Einstein describes the large scale or macroscopic shape of the gravitational field. The fundamental difference is spacetime no longer exists. Spacetime is an illusion that emerges from the gravitational field. The gravitational field sets the boundary of the cosmos. Outside the cosmos is now an empty void with no physical properties, including space and time, which cannot exist without gravity to produce the illusion of space and time we experience.

John Wheeler finds it quite instructive to think of the surface of an ocean as similar to the cosmos. The surface of an ocean is an imaginary boundary between air and water. It contains no matter. From a high altitude, it looks as smooth as glass, reflecting a clear sky. In the metaphor, this refers to the "reality" of classical physics which we experience. But as we get closer, we see it has a very complex shape as waves of different heights and directions create a somewhat chaotic surface. Theoretically, we could disassemble this random surface to discover the source of every wave. This is what quantum physics has done for the cosmos. Now Wheeler explains gravity is the magical surface where all of this activity plays out. If we look very closely, the surface reacts to the action above it, creating a fine layer of foam. It turns out the mathematics of loop theory is like that of foam, called the Wheeler-DeWitt equation. The quantum foam results from collisions between random quantum particles as they pass through and distort the gravitational field. The collisions between particles obey Newton's laws and conservation of energy. As in a bumper car arcade, higher energy particles cause most collisions. The result is the transfer of energy from higher energy to lower energy particles. Eventually, all particles will end up in the same lowest possible energy state. Thermal equilibrium at 3° above absolute zero is the predicted final state of the cosmos. Don't worry. Our universe is a vibrant youngster. It will be a few trillion years before the cosmos dies of old age.

It should be noted the random motion of particles is the source of friction, which is low frequency infrared energy, which is a source of heat. The irreversible dissipation of energy in the form of heat is called the second law of thermodynamics. Entropy is the measure of randomness or loss of usable energy. Entropy is not a property of fundamental physics. Even so, it is a secondary or derived process, which sets the direction of time. For this reason, time cannot be reversed. On the other hand, entropy is the only obstacle preventing the restoration of a perfect eternal world, as prophesied in the Bible.

The first law of thermodynamics states the total energy of a closed system, such as our universe, is constant. Recall the massive energy of a black hole cannot escape like it was a wormhole or escape route. The energy of a black hole leaks out into space as low grade thermal energy.

The idea randomness at the particle level is decreasing the availability of useful energy. When the fabric of spacetime is talked about, it can be misleading because we are talking about the quantum gravitational field.

There is a fly in the ointment; it is the probabilistic nature of quantum mechanics. It forms an impenetrable underbelly of all of physics. This eliminates the possibility of absolute precision in physics and clouds the future of physics or even portends the end of physics, according to Hawking. Equally troubling, Bertrand Russell asks we consider the set of all sets that do not contain themselves. Can this set contain itself? If it does, there is a contradiction; and if it doesn't, there is still a contradiction. This logical paradox asserts nothing physical can contain everything that exists, and there can be no such thing as a theory of everything that is assessable to finite minds.

Physics professor Eugene Wigner confirms what all physicists should believe: "It is true physics as we know it today that would not be possible without a constant recurrence of miracles similar to one of the helium atoms, which is perhaps the most striking miracle that has occurred in the course of the development of elementary quantum mechanics, but by far not the only one. The number of analogous miracles is limited, in our view, only by our willingness to go after more similar ones. Quantum mechanics had, nevertheless, many almost equally striking successes, which gave us the firm conviction it is, what we call, correct. Dr. R. G. Sachs, with whom I discussed the empirical law of epistemology, called it an article of faith of the theoretical physicist, and it is surely."

Cosmic Destiny

"The Lord will come like a thief in the night." 1 Thessalonians 5:2

We all come with an expiration date, so too the cosmos. This is unfortunate because universal laws work just fine without the time dimension. But like us, the universe is aging. The lowly hydrogen atom with only one proton and one electron is the fuel of stars, and hence the cosmos. When the fuel runs out, stars will die off and the cosmos will become a frozen wasteland in a few trillion years. In five billion years our sun, a white dwarf, will end its life in a blaze of glory called a red giant; it will incinerate the earth. The Andromeda galaxy will collide with our galaxy in 3.75 billion years; gigantic black holes out there capable of swallowing galaxies, quasars, send deadly gamma ray bursts across the cosmos. Hawking claims rogue Higgs particles could destroy the universe. There are many more immediate cataclysmic cosmic threats. Five mass extinctions have already occurred due to gigantic volcanic eruptions and asteroid collisions. The last asteroid took out the dinosaurs 65 million years ago. A 4-mile diameter meteor penetrated 12 miles into the earth along the Yucatan coastline. It left a 100-mile-wide crater, and the plume of ash and smoke plunged the earth into total darkness, causing the earth to freeze over. It is interesting to note, if the meteor hit 1 second later, it would have fallen harmlessly into the Gulf of Mexico, sparing the dinosaurs, and we would not be here today. Our planet is overdue for a cataclysmic event. But closer to home, the doomsday scenario uses statistical methods to predict the extinction of the human race when the population reaches 70 billion in about 9,000 years. There are a hundred seconds left on the 24-hour Doomsday Clock.

When ordinary stars like ours burn out, they become space junk. Black holes are the trash bins of the cosmos. They are gravitational monsters, which lunch on anything that comes their way. There is no escape from the cosmic dump. They are where everything will end up. Inside a black hole, immense gravity crushes atoms down to nothing. The density of matter in a black hole is the same as if the earth were reduced to the size of a golf ball. The black hole at the center of our galaxy contains gravitational mass comparable to the crushing 4-million suns.

The existence of black holes was first predicted in 1965 by Stephen Hawking. Since the first black hole was discovered in 1972, nearly a trillion interstellar black holes have been identified and every galaxy has a black hole at its center. Inside a black hole, reality slowly disintegrates into nothing over a few billion years. In general relativity, a black hole is an infinite singularity in need of a theory of quantum gravity, just like the creation event. In the meantime, Hawking's theory of black holes combines quantum physics with all the physical forces and thermodynamics into one elegant and straightforward equation.

As we move up the hierarchical layers of physics, new forms of complexity emerge, black holes being an extreme example. Black holes have properties that are unique and not present or predictable from the quantum level. While the behavior of an individual atom cannot be predicted, the properties of an extensive collection of atoms can be predicted by high-level patterns captured with the mathematical methods of statistics.

In the nineteenth century, Ludwig Boltzmann developed the theory of statistical mechanics and thermodynamics, which describe the properties that emerge from an extensive collection of atoms. His second law of thermodynamics describes the emergent property of entropy, which increases as the availability of useful energy in a system, including the entire cosmos, is always decreasing. Note that available energy is very different than total energy. For example, if an ice cube is placed in a glass of water, the system will proceed to thermal equilibrium. The property in question has to do with the order in a system. Boltzmann observed the melting ice cube could be explained by a loss of order as the melting ice disappears in the glass of water. The process is irreversible because the energy difference between the ice and water cannot be recovered. Because entropy is irreversible, it is the only physical property of the cosmos that sets the direction of time.

Entropy is a measure of randomness or disorder. A simple example will illustrate how entropy is used as a measure of randomness. A new deck of cards comes out of the box in perfect order. There is no disorder, so the entropy is zero. If the deck is cut, the entropy is one because it can be put back in order in one step. As the deck is shuffled, its randomness or entropy increases. Entropy is maximum when the deck is totally randomized. The maximum entropy is computed as follows: There are 52 possible cards, which could be on top of the deck. Then there are 51 cards left to occupy the second position. The total result is the series 52x51x50x49 – 4x3x2x1. This is 52 factorials, written 52! which equals 10^{16} possible arrangements or the entropy of a random deck. The probability of a random deck just happens to be in perfect order by accident is one chance out of 10^{16} or one in 1,000 trillion. Anyone who has played 52 pickup knows it is easier to randomize a deck than to put it back into perfect order. A broken egg cannot be restored. It is much easier to break toys than fix them. This is why war is so efficient.

Entropy is an emergent property because it has no relationship to the fundamental cosmic laws. Einstein proved spacetime is a single entity. We should be able to move back and forth in time, just like we do in space. If this were possible, we could live forever. Miraculously, the formation of the first living cell defied the law of entropy. Living things have the unique ability to extract energy in nature to produce a complex organization with a high degree of order. This reversal of entropy defies the law of irreversibility that operates on the rest of nature. It is as though the cosmos was originally designed for us to live forever. Then, for some reason, a monkey wrench was thrown into the works, and now we must die. "The Lord God commanded the man, saying, "Of every tree of the garden you may freely eat; but of the tree of the knowledge of good and evil you shall not eat, for in the day you eat of it you shall surely die," Genesis 2:17.

Closer to home, global warming is melting glaciers, polar ice caps, and permafrost at an alarming rate. Santa was forced to move to the South Pole. Melting ice has resulted in a 25-degree increase in Artic temperature, causing the Arctic vortex to spread to lower latitudes and changing the path of the jet stream. Firestorms, fire-tornados, never seen before in the West, now occur year-round due to extreme droughts. Hurricanes and floods in the South, tornados and droughts in the Midwest are becoming stronger and happen more often. Coral reefs, the nurseries of the ocean, are dying.

Oceans are rising faster than predicted due to the expansion of warming water. Loss of landmass will cause catastrophic famine and death, resulting in many wars.

The covid-19 pandemic has brought attention to infectious diseases that plague our planet, like bubonic plague and Spanish flu. A virus that laid dormant in the permafrost for 40,000 years came back to life due to global warming. It killed several thousand reindeer in Russia. There are thousands of pathogens waiting for their turn in the sun to attack us and our food supply. Finally, the ocean conveyor currents will stop and the earth will enter an ice age, where cities like New York will be buried under six feet of solid ice. Long before this happens, terrorists will obtain nuclear weapons and bring on Armageddon. We are nothing more than a brief epiphenomenon with all the cards stacked against us.

Jesus said, "I tell you the truth, not one stone will be left on top of another, everyone will be thrown down. The disciples asked, What will be the sign of your coming and of the end of the age? Jesus answered, watch out that no one deceives you. Nation will rise against nation. There will be famines and earthquakes in various places," Matthew 24:2-7.

Quantum Information Theory

"All matter originates and exists only by force. We must assume behind this force is the existence of information, a conscious and intelligent Mind. This Mind is the matrix of all matter." Max Planck

We have been thrust headlong into the information age. The amount of information produced every day now exceeds the total of all previous information in history. In the internet world, while no one would doubt the amount of information out there, few would consider how information shapes our mental view of reality, our lives, our bodies, and the material world we inhabit. Just like a computer, our brain is designed to process information. Pain is nothing more than information of the highest importance to our brain. Information transforms everyday reality. It moves and forms our thoughts, words, and opinions. It changes our vocabulary. It crosses both language and geographical boundaries, creating new concepts. It can frighten us. It can excite us, depress us, mislead us. Information offers new meaning to old ideas.

Information now plays the central role in modern theoretical physics. The materialistic view of the cosmos has been undercut by a deeper understanding based on mathematical laws, quantum experiments, information theory, and surprising insights. The universe is no longer the mindless machine of the Victorian era. The cosmos is now the thought patterns of a great Mind, creating the illusion of reality we experience. This reality has been reduced to its most fundamental elements; the information carriers called quantum waves. Quanta contains instructions from the creation event, which work together to evolve the universe from nothing. This latest contender for a theory of everything has been called the Creator's Voice. In this theory, reality is a multifaceted communication network, delivering a world of information to the mind of humans sent by the Mind of God during creation. This strongly suggests any proposed theory of everything will be incomplete until it includes a mathematical theory of mind, which resolves uncertainty in quantum theory, starting with the observer effect. Einstein was correct when he told Bohr, "God does not play dice."

The four cosmic forces are now understood as four quantum wave fields: Higgs ocean, Dirac sea, gravity, and EM fields. These ubiquitous quantum fields are the foundation of the cosmos. Each field supports the production of its particular quantum particle: Higgs boson, electron, graviton, and photon. These localized quantum waves emerge near maxima in their respective quantum field. The exchange of information by these quantum wave packets, which create what we experience as reality. The Higgs interacts with gravitons to give the property of mass to quarks. Quarks combine to form protons and neutrons of the nucleus. EM connects electrons to the nucleus to form atoms. Electrons create photons of light. Quantum fields work together by exchanging information, much like humans use light waves and sound waves to communicate. What has been described is a simplified tip of an iceberg, the size of which can never be known by finite minds.

INFORMATION EXCHANGE

The information carried by quantum waves has familiar names: energy, mass, charge, and spin. The names only have meaning in terms of how the different quanta interact with each other. The quanta can only be characterized as particles in terms of their interactions with conscious minds, the observer effect.

Conservation of information is now recognized as the most fundamental law of physics. Burning a book doesn't destroy the quantum information from which the book is made; it just increases its entropy by scrambling its information. Conservation of information means states within quantum fields can change, but the quanta themselves cannot be created nor destroyed. As a quantum state evolves, it must remain uniquely distinguishable from all other quantum states. This means all the information quanta in the cosmos yesterday, today, and forevermore were present at the moment of the creation event. The universe as understood by physics is a structure built upon the flow of information.

In 1949 Claude Shannon, a theoretical mathematician at Bell Labs, published "A Mathematical Theory of Communication." In it, he developed a fundamental connection between entropy and information. The Shannon definition of entropy is a quantitative measure of the degradation of information in a communication or information system. Correctly ordered, unambiguous information has zero entropy. Information loss due to errors or noise can then be quantified in terms of increased entropy.

Similarly, entropy can be associated with the uncertainty of our knowledge of the structure of a system such as a cosmos, giving a precise, mathematical meaning for information uncertainty as it corresponds to physical disorder. One of the most fundamental properties of the universe is a direct result of the second law, of thermodynamics. According to this law the entropy of the cosmos began in a perfectly ordered state of zero entropy and will eventually reach a maximum value when all usable information and energy have been captured by black holes where it will all spit out as useless information garbage over trillions of years. The Penrose number estimates the amount of information quanta in the cosmos and the entropy of the cosmos when all useful information is used up.

The connection between entropy and information is fertile ground for study in physical and biological sciences. For example, the lower entropy of the past is directly related to our ability to remember the past. Increasing entropy is directly related to the degradation of information. The information needed to make a human is contained in the genome, consisting of three billion bits of information. A human ovum has zero entropy while increasing entropy, is the aging process.

How entropy continually increases while information is conserved in the cosmos is wrapped in the problem of quantum gravity. It seems possible the hidden information of the cosmic code is entrenched in the invisible dimensions of string theory. If this hypothesis is correct, this inaccessible information may be more fundamental than the information of observable quantum states. Physicist and Hebrew scholar, Gerald Schroeder, suggests the six extra dimensions of string theory are God's domain of influence. Ancient Rabbi Maimonides noted there are ten dimensions of reality revealed by the ten times the phrase "and God said" is recorded in the first chapter of Genesis. [31]

Modern theoretical physics allows for the creation and destruction of matter but does not allow for the creation or destruction of energy or quantum information. In 1996 Professor of physics at Stanford, Leonard Susskind recognized this fundamental principle of physics was violated by Hawking's theory of black holes, which claimed information quanta disintegrate and cease to exist as black holes decay. Hawking spent ten years reanalyzing his equations. He finally agreed information quanta within black holes cannot be destroyed. A scientific theory can now explain the workings of the cosmos quite well, except for two problems, how it was created and how it will end. At these endpoints of the universe, namely the big bang and black holes, Einstein's equation reaches a singularity, i.e. goes to infinity or blows up. Physics ends long before infinity begins. The largest number you can imagine does not begin to describe where infinity begins. Science is stuck at the mysterious, unknowable, infinite singularities of the equations and the inability to observe any physical phenomena on their other side.

No matter what lies outside our little pocket cosmos, scientists have found something quite profound about it. The ability of the human mind to receive cosmic information means we can infer the source of information is an Intelligent Mind outside the universe. The source of material reality must be the God of Genesis, who spoke the cosmos into existence. This is the foundational revelation of monotheism. It is the God Hypothesis "smoking gun." Any scientific theory cannot refute this conclusion. No one has ever seen an equation get up and walk. The laws of physics are the word of God made manifest in our minds and hearts. There is no escape from this simple truth.

Physicist Hans Pagels wrote, "If we define a vacuum as the state for which all physically conserved quantities are zero, then we find, amazingly, the entire universe could be equivalent to nothing... Suppose, for the sake of argument; scientists find the master law of the universe, the basic software of the cosmic computer. By viewing the laws of the universe as the software, I foresee the possible future merger of physics with information." [25]

Imagine an interactive computer simulation of the cosmos. Everything on the computer screen looks and acts just like the real thing. The human mind can readily see a 3D world projected onto the 2D display. The surprise is reality is the same kind of illusion. What we perceive as material existence is a 4D illusion. Recall, the only reason a big rock seems real is that it hits you back with the same force. You created the force, not the rock. The stone is just protecting its space. Two objects are not allowed to occupy the same place at the same time. So, respect the rock's right to have some space it can call its own. You can't even see the rock. All you can see is the color of light it reflects because it didn't want to let it inside. One is led to believe there must be an inaccessible reality, which is nothing at all like what we experience. Could it not be the metaphysical reality of Plato or the Kingdom of God?

Scientific reasoning is ending at the Creator's doorstep. Science is not about proving whether or not God exists, but the evidence supports the need for an infinite, all-powerful Creator. The magnificence of the universe provokes a sense of a transcendent, eternal Being. Werner Heisenberg shared this interchange: "Wolfgang Pauli asked me quite unexpectedly, "Do you believe in a personal God?" "May I rephrase your question, I asked. I should prefer the following formulation: Can you, or anyone else, reach the middle order of things or events, whose existence seems beyond doubt, as directly as you can reach the soul of another human being. I am using the term "soul" quite deliberately so as not to be misunderstood. If you put your question like that, I would say yes."

Quantum Entanglement

10/9/2012 CNN: "Serge Haroche of France and David Wineland of the U. S. Shared the 2012 Nobel laureate in physics. The two winners have carried out remarkable experiments that have confirmed phenomena such as quantum entanglement. Their teams have shown just how strange the quantum world is and opened up the potential for new technologies such as quantum computing."

In 1964, Irish physicist John Bell developed a theory of a most mysterious phenomenon called quantum entanglement. This theory predicts quantum information (i.e., the exact state) of an atom, electron, or photon in one place can be instantaneously linked with another particle's state at any other site in the universe, defying the speed of light. Einstein called this "spooky physics."

In 1997, Nicolas Gisin at the University of Geneva conducted an experiment that validated quantum entanglement is, in fact, a real phenomenon. It is called an unconscious property because it violates the speed of light. Both quantum information theory and quantum entanglement are non-physical connections of information between quantum particles. As bizarre as it seems, this theory does not violate any known laws of physics since the particles always exchange information faster than the speed of light and never pass through the speed of light barrier. Physicist John Wheeler states, "In quantum physics, the concept of time-dependent causality is vague, even lost."

Based on their prior history, two particles can become causally entangled such that the observation of either particle's state will be transmitted instantaneously to the other particle and cause the other particle's state to become the same no matter how far their separation in the universe. For obvious reasons, quantum entanglement is called a non-locality property. It challenges a fundamental assumption of spacetime physics. Scientists used to think a fundamental property of space and time is it separates and distinguishes one object from another. Quantum physics radically challenges this view. No amount of space weakens quantum entanglement. Even spatially distinct objects can become entangled in our minds as demonstrated by the Necker cubes.

Some physicists like Professor Fred Wolf [32] speculate quantum entanglement connects us all in one universal whole because everything in the universe has emerged from one source, the creation event. The non-locality property is now accepted as an inherent property of quantum physics. The state of the world or any subsystem of the universe is unknowable because the state of each quantum particle is unknowable and because their quantum fields interconnect all particles in the universe. An idea closely related to non-locality is the possibility of hidden dimensions beyond our common four dimensions of spacetime, which are predicted by string theory. The inability to predict the precise outcome of quantum experiments strongly suggests nature is still withholding information from us.

Quantum physicist Antoine Suarez suggests God's role in human affairs may be hidden in quantum entanglement. In what is called the before-before experiment, Suarez and others found whatever causes twin photons to act in tandem; it made no difference in what order the photons were created. The causality principle of physics can be violated. "There is no story that can be told within the framework of spacetime that can explain how these quantum correlations keep occurring," Suarez says. "These results have intriguing philosophical implications," he notes, "especially for the spiritually inclined. You could say the experiment shows spacetime does not contain all the intelligent entities acting in the world because something outside of time is coordinating the entanglement of elementary particles. Information is being passed around the universe subliminally with total disregard to causality. Physics experiments cannot demonstrate the existence of God, but this test shows today's physics is compatible with Western religious traditions. There is strong experimental evidence for accepting nonmaterial beings acting in the world."

Jean Stained, a mathematical physicist, says quantum entanglement "demonstrates the existence of a level of reality that escapes time, space, energy, and matter, yet still has a causal effect on our material level of reality. Science has suggested through quantum physics that it alone cannot provide a complete picture of reality. It has provided the basis for a credible way to understand the existence of God because the world no longer limits itself to our level of reality." He asserts quantum physics refutes materialism and suggests a way to think about the existence of God. This may be a pitfall John Polkinghorne wants to avoid, "If you explain away every scientific unknown by invoking God, you end up with the God-of-he-gaps, one can be eroded anew every time a new part of the scientific puzzle is solved." [33] Christopher Isham, a theoretical physicist, adds, "The trouble is if science later advances, God will be left high and dry." Isham questions the merit of trying to validate religious experience by appealing to science and risking the God-of-the-gaps embarrassment. "For me, religious belief is more about mystical feelings about the world, and God is something one encounters in one's self," says Isham. "And if scientists ever did manage to uncover mechanisms used by God to influence the physical world, it would become even harder to defend why God does not use this power to alleviate suffering. It does rather raise the question of why the universe is, frankly, a bit crummy." Is it not true our creature nature is selfish? If life were perfect, would we find a need for God? Of course, we can conceive of a perfect world. Has not God promised a perfect world? God wants to teach us his ways, not destroy us. "The way I see it, if two objects touch, can somehow then communicate data with each other instantly, no matter how far apart they are brought, through quantum entanglement, then surely if there is a God, then He communicates with us on some level, no matter how vast the distance we are apart." [25] Hans Pagels

Reflection

"A careful analysis of the process of observation in atomic physics has shown the subatomic particles have no meaning as isolated entities but can only be understood as interconnections between the preparation of an experiment and the subsequent measurement. Quantum theory thus reveals a basic oneness of the universe. It shows we cannot decompose the world into independently existing smallest units. As we penetrate matter, nature does not show us any isolated basic building blocks, but rather appears as a complicated web of relations between the various parts of the whole."

Frito Capra, "The Tao of Physics, On Quantum Theory"

Nature does not give up its secrets easily. The quest for scientific knowledge is like wandering in a dark forest. Now we know space and time are not as real as they seem. Instead, they are a subtle illusion created by gravity giving physicists a tantalizing clue gravity itself may be hiding even more profound secrets yet to be discovered. After all, gravity is the greatest obstacle to becoming the masters of the universe.

Immanuel Kant suggested it would be not merely impossible to do away with space and time when thinking about and describing the universe; it would be impossible. Nevertheless, many of today's leading physicists believe space and time are not as fundamental as they appear. Just as the hardness of a rock emerges from the collective properties of its atoms, and the smell of a flower emerges from the collective properties of its molecules, and the flight of birds arises from the collective properties of their anatomy, so too, the properties of space and time emerge from gravity, and gravity controls the motion of free energy like light and condensed energy like matter. Now the question is, can physicists figure out how to control gravity? Did not Jesus demonstrate to us that He could defy gravity? Physicists with a great imagination, like Einstein, never stop speculating about the future of science. It is a worthwhile and interesting exercise. It places our current undertaking in a broader context and emphasizes the larger goals of science. When speculation turns to the future of spacetime itself, it takes on a mystical quality. We're considering the nature of the very essence of our sense of reality. There is no question regardless of future discoveries, space and time will continue to dominate our experience. Space and time, as far as everyday life is concerned, are here to stay. The scientific understanding of the deeper layers of reality goes way beyond the reality of everyday familiarity. We may understand our little planet well enough, but the universe is an inhospitable stranger. Modern physics has delved deeply into the mysteries of the cosmos, which have remained hidden until the last century. But we still know very little about the dominant constituents of the cosmos, dark matter, and dark energy. We know 72% of energy is dark energy, and 83% of all matter in the universe is dark matter. In the future, the ability to detect gravity waves holds the promise of deeper cosmology and may one day allow us to travel on intergalactic gravity waves and see beyond the confines of our cosmos.

Existence Revisited

"Man, beginning with himself, can define the philosophical problem of existence, but he cannot generate from himself the answer to the problem.

The answer to the problem is the infinite, personal God is there, and He is not silent." [14] Theological teacher Francis Schaeffer

"No one can find out the work God does from beginning to end." Ecclesiastes 3:11

Physics has answered the question "why there is something rather than nothing" by discovering there is no fundamental difference between the two. So, the better question is, "Why is there existence rather than non-existence? What does it mean to exist? The very existence of existence is beyond comprehension. There is no rational explanation for existence. It just is. Science cannot prove or disprove the how or why of existence. What existence means is the realm of God, not science. We just are here to wonder about the how and why of existence.

In philosophy, the study of reality and the nature of entities, which are said to exist, is called ontology. The question is whether or not physical reality as we perceive it is the framework for all that exists. Unfortunately, there are no first principles, logical reasons, or scientific theories which can explain existence. For anything to exist, it must be assumed some prior something must have always existed. There is no scientific evidence that existence can create itself from non-existence, nor is there a shred of scientific evidence physical existence is in any way infinite or eternal. Why then do we exist when reality appears to be a finite illusion? The source of existence must in some way transcend the finite illusion of material existence. Scripture has provided a transcendent framework, which is a viable answer. Only God can explain himself. We cannot truly know God by his creation any more than we can know a watchmaker by examining his watches. But we can recognize the intelligence and skill of a watchmaker and God. Theological philosophers have argued what we can understand as reality only exists in the mind of God. This is the illusion of physical reality. By this argument, God is the infinite Creator and omniscient Observer who provides the framework of existence, for without a transcendent source called God there can be no existence. This is the core belief of the God Hypothesis.

"To be or not to be? Is the question." Atheism rests on the finality of "not—being." Reductive materialism is the belief the only things that exist are transitory material objects, and all phenomena, including consciousness, are the result of material interactions. Science left Plato's metaphysical "essence of being" behind. Philosopher William James observed, "From nothing to being, there is no logical bridge." Even without God, the uncertainty of existence and the question of being or not-being remains. Psychologist Carl Jung answers as follows, "As far as we can discern, the sole purpose of human existence is to kindle a light of meaning in the darkness of mere being." The logical basis of materialism is inadequate and superficial because it is not a framework that explains how or why anything exists. It is merely the failed hypothesis of the rationalism of the nineteenth century, which presupposes God does not exist.

Reconciling reality with our perceptions of reality is an important philosophical issue of the past. With or without God, the uncertainty of existence obscures the question of being or not-being. Being is more than atoms. Science left the importance of "being" behind. What it means to be a human being is not contained in the methods of science. Whereas atheism is the ontology of not-being, the ontology of being is conscious awareness of the timelessness of being.

Kant realized believing in timeless being must precede knowing it. Pascal and Descartes knew finding truth is an act of reason and faith. Pascal argued existence cannot be proved until it is first believed. Descartes believed the "it" that makes us "beings" is the riddle of life. He pointed out, "I do not exist," is self-contradictory. The act of saying one does not exist assumes someone must be making the statement. It is therefore, reasonable to conclude conscious beings know they exist as real beings, while all else in the cosmos is an uncertain illusion.

Descartes' unexceptional claim, "I think therefore I exist," contains within it the essence of a paradox that runs throughout the history of human thought. Thinking is a time dependent process; existence is a time independent state of being. When I think, my mental state changes with time. But the being to which the mental state refers remains the same. This is an ancient metaphysical problem, which has renewed interest in modern science. Objects in the world are constant, just like our identity, yet like us; they are always changing. Stuff happens. The present becomes the past, and the future becomes the present. Existence is this paradoxical conjunction of being and becoming. Humans are the only creatures aware of becoming.

Does the future portend hope or despair? Our knowledge of mortality seeks enduring aspects of existence. People come and go, trees grow and die, even mountains gradually erode, and we now know the sun cannot keep burning forever. Is there anything that is genuinely permanent? Man, once thought celestial objects, the sun, moon, planets, and stars were eternal fixtures of the cosmos. But we now know better. We now know whatever is constant or eternal does not exist in the material world. Only God gives us hope of permanent existence. But then the paradox of existence confronts us. How can the changing world of our experience be founded on the unchanging world of an abstract concept like God?

At the dawn of civilization, the Israelites had formed a concept of God who created a finite temporal universe and yet exists independently of the universe as an eternal Being. Greek philosophers also confronted this dichotomy of reality. For Plato, true reality belonged in the transcendent world of unchanging abstract "ideas and forms," a domain of pure mathematics and geometry. This was the realm of pure being, inaccessible to the perception of our limited senses. For the Jewish prophets and Greek philosophers, the changing world of our direct experience is a fleeting illusion. The material world was relegated to a pale shadow of the reality beyond our perception. These great thinkers recognized a fundamental tension between being and becoming, between the eternal reality of being and the changing world of experience. The Greek culture failed to penetrate the paradox of reality because their pantheon of mythological gods did not go beyond the limited reality of the temporal world. In stark contrast, the Israelites conceived of an infinite, all-powerful God capable of creating eternal beings. The Genesis creation story is the most perceptive religious explanation for existence ever written. It matters little, whether it is literally true or a deeper revelation. The meaning, significance, and intent are transparent and unchanging.

Let me repeat: The question from which everything else follows is "Why is there existence in the first place?" A finite cosmos is direct evidence there must be a transcendent infinite existence. As we consider the finite aspects of the world we know surrounds us, the limited nature of spacetime and matter from which we are constructed, the answer is undoubtedly yes. The finite and the infinite cannot have the same nature. A superior Being, above or outside of the material world must have preceded the cosmos or has the cosmos embedded within it. We are so much a part of existence we take it for granted. Existence is an unexplainable given in science. In science, existence is assumed true, but its meaning is becoming less clear in scientific terms, certainly not clearer. We are uniquely real and different from the inanimate world in which we live. We must form a belief in the existence of the Creator according to our life experiences.

Science and religion seek to explain the source of existence and the power, which sustains life and the cosmos. They go together like birds of a feather when we see through the limits of mankind's knowledge to the limitless source of all that exists. Scripture begins with the story of creation. If we want real purpose in our lives, we must believe the universe was created for a purpose; otherwise, life has no meaning beyond simple existence, just like pond scum. Revelation in Scripture and scientific knowledge of nature are two aspects of one creation. "God is more easily found in the works of nature than in the works of man." Ralph Waldo Emerson, philosopher. Unfortunately, the idea that the study of nature might have something to add to our understanding of spiritual existence continues to be anathema to most of the religious establishment.

Classical physics makes sense to most people. It forms the basis for our collective understanding of the natural world. But modern physics is a totally different story. It is incomprehensible to most people. And for a good reason. It defies common sense, just like God. Science has now discovered the cosmic code, which is non-physical information that is more fundamental than time, space, energy, or matter. In the words of John Wheeler, the renowned former President of the American Physical Society, recipient of the Einstein Award, and Princeton Professor of physics, "underlying all existence is an idea, the bit of information that gives rise to the "it" of matter." Gerald Schroeder wrote, "The substructure of all existence, we suddenly realize, is ethereal, an idea, wisdom, truth, or in Hebrew, *emet,* an all-encompassing reality. Emet is the ultimate building block from which all we see and feel is constructed. Just as the essence of all matter is something as ethereal as energy, as per Einstein's fantastic insight, so the primary essence of energy is an even still more elusive cosmic code. Existence is the universal expression of an idea, an eternal consciousness made tangible. We are the idea of God. If we can discover that idea, we will have ascertained not only the basis for the unity that underlies all existence, but more importantly, the source of that unity. We will have encountered the mind of God." [32]

There is something about nature that is much more striking and inexplicable than its design. Without inductive reasoning, we could not learn from experience, memory would be unreliable, language would be useless, our very existence would be no better than that of the lowliest creatures. It would indeed be true that we would be nothing more than "chemical scum." Inductive reasoning is so ingrained in our everyday experience we don't even think about it. But what if we never knew when the sun would rise or water freeze. Great philosophical thinkers such as David Hume and Bertrand Russell found this assumption about nature inexplicable. Why is there natural order exhibited in a world we know was created out of apparent chaos? And moreover, we haven't the slightest rational justification for assuming it will continue tomorrow. If we believe the future will always be like the past when it becomes the past, Hume and Russell would point out, you are assuming the very thing you are trying to prove. Science can only take the continuing regularity of nature on faith as Einstein said, "what is so incomprehensible about the universe is that it appears to be comprehensible." If one asks for a rational explanation for existence, then we have no choice but to seek that explanation in something beyond the physical world, in something metaphysical because a physical cosmos cannot contain within itself an explanation for itself. What sort of metaphysical agent might be able to create a universe? It is not sufficient to naively invoke a God instantly creating the cosmos. Richard Dawkins makes the following cogent argument:

"The bald statement God created the universe fails to simplify our view of the world. A Creator introduces an additional complicating feature without explanation. It provides no real explanation unless it is accompanied by a detailed mechanism. One wants to know, for example, what properties to assign this God, and precisely how He goes about creating the universe, why the universe has the form it does. There is no way we can test the hypothesis experimentally. In short, unless you can either provide evidence in some other way that such a God exists, or else give a detailed account of how He made the universe that even an atheist like me would regard as deeper, simpler, and more satisfying, I see no reason to believe in such a Being. I would rather accept the existence of the universe as a brute fact than accept God as a brute fact. After all, there has to be a universe for us to be here to argue about these things." In other words, Dawkins believes we can accept creation without explanation or we can seek a plausible explanation through science. If the scientific explanation is consistent with the creation story in Genesis, then not only the credibility of Genesis is established, but also the credibility of all Scripture. We now have a detailed, timeless notion of the Creator, one who is an infinite, all-powerful Being with a Mind and a purpose for creation. This Creator must be responsible for all the laws of nature. Scholars now recognize that science arose in Western civilization because of the monotheistic Western religious belief in an all-powerful, personal God who created and sustains an orderly universe. As proof for the existence of God, the regularities of nature are inescapable. You can always say we don't know why things are as they are. But it is inescapable evidence for the God of the Bible.

There will always be more unknowns than knowns. There will still be unknowable unknowns. Unknowns become knowns in the least expected ways. Our mind is our only connection to reality. The reality we experience only exists as an illusion in our collective minds. How mind emerges from illusion of matter is powerful evidence for the mind of God. Our minds share an illusion created by God, and God has been programmed into our minds by evolution. Some people claim it is possible humans evolved to think about God without the need for God to be real. Belief in a Creator is a more compelling and satisfactory explanation for existence than imaginary explanations created in the mind.

The spiritual world of our existence can only be known by divine revelation. Divine revelation does not contradict external knowledge because it is understood as internal knowledge. Religious experience is personal, like real love that is too deep for words. Many such experiences can only be acquired and understood by your desire for the experience and the willingness to risk embarrassment, even failure, to achieve it.

Plato recognized the idea of an object in our mind is different from any particular object. Even if we kick an object, the pain we experience is all in our heads. We don't experience objects; we only experience our mental perception of objects. In Kant's philosophy, the ultimate reality of an object, the thing-in-itself, can be conceived in thought, but cannot be perceived as inexperience. Our senses are our only means of perceiving reality.

Furthermore, information in the brain can only be stored and retrieved as word symbols, a superficial representation of reality. What then can we know and not know about reality? The world we know as physical reality is the material world of physical phenomena we can understand only by empirical science. It is limited to what can be identified by sensory perception. Kant and Plato concluded objective reality is invisible, unknowable by empirical science. Objective reality exists behind the veil of human experience, just as sound is unknowable to the deaf. Kant and Plato captured a vision of the mystical quantum world in their ideas.

Our perception of it limits our knowledge of reality. Kant is not claiming our perception of reality is unreliable. He is just claiming our perception of reality can never penetrate subjective reality to know what objective reality is. Objective reality is hidden from us. Our perception of reality is just a symbolic copy of subjective reality. Our mind learns to create a virtual reality that may not be there. Sound, light, color, smell, and taste do not exist as physical properties of reality. Our mind develops these senses to make sense of reality. How can we compare our perception of reality with reality itself when all we have is our perception of reality? If we believe our perception of reality is the same as actual reality, then we have just made a giant leap of faith.

This picture is a simple example of how perception can be deceiving.

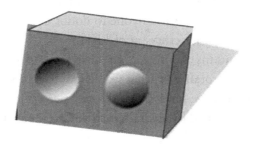

The shaded discs are identical except one is lighter on top while the other is lighter on the bottom. Experience has programmed our minds to assume light comes from above, so we cannot help but see a bulging disk next to a concave disk. Although you know your perception is being tricked, there is nothing you can do about it. Artists have many such rules to help create depth on a flat surface. For example, lines converging at a distance is another rule in our brains to imitate distance. Human recollection of the past is notoriously unreliable. The reality we experience is just one illusion after another. We are not designed to experience the truth about reality. We literally would not be able to handle it.

Kant's purpose is to demonstrate there are limits to what can be known through science and reason. He is not intending to devalue science or reason. On the contrary, he is a firm advocate of science and reason as long as it is recognized they have their limits. He wants the skeptic to recognize the fact that there is more to reality than just our experience of it. Kant's argument seems to defy common sense, but it is so powerfully rational it is irrational to reject it. What could there be about reality that is so important we have to suspend science and reason to know it. Kant has a good idea of what transcends reality because he was a theologian. All religions of the world benefit from Kant's brilliant, irrefutable arguments. Kant has proven on the basis of reason alone that belief is an essential component of reason. In his own words, "We now have room for faith." Kant has revealed atheism to be arrogant in the delusion that reason is the only legitimate source of knowledge. It is not religious believers who are arrogant. They can now humbly admit the limits of human knowledge, knowing there is a reality greater than that which our senses or our minds can ever apprehend. "It always remains a scandal of philosophy and universal human reason that the existence of things outside us should have to be assumed merely on faith, and that if it occurs to anyone to doubt it, we should be unable to answer him with a satisfactory proof." Kant

Our mind contains within it an abstract virtual world that is constructed from the totality of what we believe to be true. We believe in the real world because we have direct sensory perception of it. That real world is an illusion created by abstract mathematical equations embedded in a cosmic code. Let me repeat. The spiritual world can only be known by divine revelation. The Bible says, "Seek and you shall find." Divine revelation does not contradict external knowledge if it is understood as internal knowledge. Spiritual experience is so unique to oneself as to never be experienced or agreed upon by anyone else. It's like riding a bicycle. It can only be acquired by your own desire for the experience and the willingness to risk embarrassment, even failure, to know it. Here is a notional diagram of different realities of existence as I envision them.

Cosmic Accidents

"God did not create the universe," Stephen Hawking.

Until a half-century, ago physicists were content in believing the cosmos was infinite or unbounded in time and space. The big bang put a big dent in that theory. We now know the cosmos is finite in both time and space. This allows the possibility of a Creator. Rejecting God, scientists who are atheists like Hawking believe the cosmos invented itself. This assertion is exactly why science is seen by many as antireligion or some kind of false religion. The question is, can we observe everything that exists? As an atheist, Hawking answers in the affirmative. As a scientist, he can't know the answer. The next question we might ask is, are we hereby cosmic purpose or cosmic accident? Hawking claims he has scientific proof we are here by accident. He says we are only so much "chemical scum."

There is no scientific theory of how non-existence could cause existence. Existence did not just appear from nothing. Existence means there is stuff. Gravity is stuff. An equation is stuff. Stuff must be created. Where did stuff come from? Science doesn't have a clue.

Never lost for words, Richard Dawkins posed the following challenge: "The God question is not in principle and forever outside of science. Science can make at least probabilistic inroads into the territory of agnosticism."

Dawkins has not come up with any probabilities to back up his claims. So, let's do a little math. The probability that God does not exist is the probability that everything that exists can be explained by a sequence of mindless random accidents, starting from non-existence and ending in the present. Existence itself has no observable properties; it is simply an a priori assumption necessary for science to proceed.

- Professor Murray Gell Mann made the following calculation, "If the big bang were a purely random event, the probability of the universe containing within itself the knowledge necessary to produce life is estimated to be one chance in 10^{229}." To put this in perspective, suppose each of the seven billion people on the planet entered a lottery. In order to match Gell Mann's odds, someone would have to win over 24 lotteries in a row.
- Professor Steven Weinberg, 1979 Nobel in physics, wrote, "Life as we know it would be impossible if any one of several physical quantities had slightly different values." [23] He quantified the fine tuning of the big bang to be one part in 10^{120}.
- Professor and Fermi Lab astrophysicist Michael Turner commented wryly on fine tuning, "The precision is as if one could throw a dart across the entire universe and hit a bullseye one millimeter in diameter on the other side."
- Caltech physicist and astronomer, professor Hugh Ross, made the following calculation: "The probability for occurrence of 159 parameters necessary for human life is one chance in 10^{412}. The number of planets in the observable universe is 10^{24}. Thus, there is one chance in 10^{388} that life on our particular planet and no other planet occurred by pure accident.
- Considering only seven scientific facts about our Goldilocks cosmos, astronomer Lee Smolin computed the probability of accidental existence of the cosmos from the big bang to be one chance in 10^{250}. It is certainly a truth of a scientific kind to say we are not here by accident.
- Sir Fred Hoyle, cosmologist and mathematician at Cambridge, estimated the odds of a living cell emerging from inanimate dust to be one chance in $10^{40,000}$. There are 10^{80} atoms in the cosmos.

In other words, the probability of life beginning by natural causes is way too small to ever be explained by purely random chance. Even if we discover life on another planet, it would not explain how life began. Hoyle rejected his belief in atheism. He wrote, "A commonsense interpretation of the facts suggests that a Super Intellect has monkeyed with physics, as well as chemistry, and biology, and that there are no blind forces worth speaking about in nature. The numbers one calculates from the facts seem to me so overwhelming as to put this conclusion almost beyond question." "The Intelligent universe" [34] Sir Fred Hoyle

Biologist Michael Behe contends the living cell could not have appeared by natural causes because it has irreducible complexity. [35] That is, the mechanisms within a living cell cannot emerge or function independently of the whole cellular structure, just as a Swiss watch mechanism or a mouse trap is irreducible.

The simple fact of entropy is irreversible proves universes cannot be created by accident. Once a universe exhausts all its available energy, it is dead forever. All this business about universes being born and dying, then being born again, is hogwash.

The cosmos could not be the result of accidents because the odds are somewhere between infinitesimally small and zero. Without an infinite cosmos, many new ideas popped up: recycling universes, parallel universes, stacking-doll universes, a multiverse, and now an infinite megaverse. The probability of a particular universe such as ours is then the probability that our universe exists by chance among the totality of all possible universes. To ensure this result, some cosmologists assume the laws of the cosmos apply to whatever exists outside our cosmos without any evidence. Then they assume the 20 fundamental constants of physical laws can assume any possible value, resulting in 10^{500} possible universes. Such calculations are known as "forcing the argument" because one can prove anything one wants this way. It's like proving that a monkey will type Hamlet if 10^{500} monkeys are sitting at a typewriter. Unfortunately, a finite number of universes still leaves an infinite gap problem.

Infinity is back on the scientific front burner. Atheism -of-the-infinite-gap came to the rescue. The idea of an infinite number of universes from three centuries ago was resurrected. Leibniz wrote, "There is an infinite number of possible universes, and as only one of them can be actual, there must be sufficient reason for the choice of God, which leads Him to decide upon one rather than another. And this reason can be found only in the fitness, or the degree of perfection which these worlds possess." Voltaire satirized Leibniz as professor Pangloss in the novel "Candide." Candide searched all the worlds to finally conclude that Dr. Pangloss was right; "This is the best of all possible worlds."

Now our cosmos is just a "Johnny-come-lately." This is the newest hypothesis put forward by Hawking and others. An infinite number of universes is just the ticket. With an infinite number of universes, anything can be proved with certainty. There is one problem; such a theory can't be proved or falsified. It is not science; it's just a scientific leap of faith. Lawrence Krauss stated, "Almost every logical possibility we can imagine extending the laws of physics as we know them, on small scales, to a more complex theory, suggests that, on large scales, our universe is not unique." [36]

In 2010, physicist turned atheist, Sean Carroll, outed himself in the book, "From Eternity to Here." [37] It is not a travel guide to heaven. Instead, it explains the metaphysics of infinity using the technique of quantum hand waving. This is nothing more than atheism-of-the-infinite gap. which makes no sense because nothing made of matter can get from eternity to here. Such claims lack empirical credibility. Something physical, starting at infinity, could never get here! Even the smallest possible fraction of infinity is still infinite. Infinity was banished from physics a century ago.

The fact is, we live in a finite time warp with no escape. One must ask, why are atheists afraid to admit what they don't know is infinite compared to what they do know?

It is not possible to logically support infinite chains of random events to explain the cosmos. Even if such a pure luck hypothesis were coherent, which it is not, it would still be wrong because it cannot be tested empirically. A theory of infinities is a theory that anything and everything will happen. Infinities have always been and always will be the bane of physics. An infinite number of universes is inconsistent with the theory that all universes have a finite beginning. An infinite megaverse is touted for public consumption by Stephen Hawking, Leonard Mlodinow, Sean Carroll, Brian Greene, Neil deGrasse Tyson, Lawrence Krauss, Daniel Dennett, Jerry Coyne, Richard Dawkins, and many more scientists. They don't seem to understand that infinity is a mathematical construct beyond the reach of science.

It is usually assumed there are only two possibilities: God is needed to explain existence, or existence is the result of an uncountable number of cosmic accidents. If there are no other possibilities, then the probability of these two possibilities must add up to one. Since the probability of existence by cosmic accident is virtually zero, it is virtually certain that God is needed to explain existence. The God Hypothesis is thus proved to be a scientific theory or even fact. Physicist Freeman Dyson had this to say, "I conclude from the existence of these accidents of physics and astronomy that the universe is an unexpectedly hospitable place for living creatures to make their home. Being a scientist trained in the habits of thoughts and language of the twentieth century, rather than the eighteenth, I do not claim the architecture of the universe proves the existence of God. I only claim that the architecture of the universe is consistent with the hypothesis that Mind plays an essential role in its functioning." [1] The odds are so staggeringly high against accidental existence that even talking about probability is beyond rational comprehension. Sherlock Holmes explains, "Once you eliminate the impossible, whatever remains, no matter how improbable, the truth must be."

Some argue that randomness is so inherent in nature, that uncertainty precludes the credibility of sweeping conclusions like the existence of God. In fact the randomness we experience is an illusion. The chaotic events we experience, like the weather and sunspots, are subjective because they are subject to the laws of statistical mechanics. If an experiment is properly instrumented, there is no inherent reason why the laws of physics cannot predict the outcome of flipping a coin. The only truly objective randomness exists at the quantum level. Perfect randomness is a startling feature of quantum physics. Somehow, all traces of causality disappear. Physicists seem to have uncovered an impenetrable barrier. It seems like physics has found the fence around God's estate.

Einstein said, "Coincidence is God's way of remaining anonymous." God is so clever, He hides behind dice throwing, just like casino operators who count their money in back rooms. It is also interesting that the title of Hawking's book is "The Grand Design" [38] and yet his book is trying to explain how cosmic accidents create universes without any need for a grand design, let alone a Grand Designer. Hawking ignores the significance of creation by believing material reality is somehow eternal, the creation process is uniform, symmetric, harmonious, comprehensible, majestic and immutable. Yet most scientists say this is just the way it is, a fluke of nature. They deny the reality of a Creator whose creation they are learning about in intricate and breathtaking detail. It all comes down to the question of whether or not science can explain existence, right down to the first quantum fluctuation which created itself out of non-existence.

2004 Nobel-physicist Frank Walczak is often quoted when he was asked why there is something rather than nothing, "Nothing is unstable. We can also show that the laws of physics are just what they should be if the universe came from nothing." Creation out of nothingness is precisely creatio ex nihilo. No scientific theory can change this fact.

Consider the scientific "lucky-by-accident" conjecture called the "megaverse hypothesis," based on the following assumptions; (1) the megaverse is infinite in space and time, (2) the number of universes is not known, but is most likely also infinite, (3) the universes are all different, but ours should be typical of a large number, (4) the laws of our universe apply to the entire megaverse.

The only way cosmologist can explain the very instant of creation is to "assume" (ouch) the physical conditions before the universe began are like those when the world will end. The universe will continue to expand at the speed of light until it completely disintegrates, reaching a zero-point vacuum energy steady state at 3°K. It is assumed the amorphous void acts something like the surface of a pot of boiling water. Over trillions of years, trillions of potential universes are formed like bubbles in the steam, then fall back into the boiling water. Over trillions and trillions of years, there is a finite probability a bubble will reach a stable state where a universe like ours will suddenly pop out. Another way to think about this is to consider that all information necessary to create a universe is distributed in an amorphous vacuum of bits of information called quanta. These quanta of information move about randomly, like the molecules in a room. Over an immense amount of time, there is a finite probability that all molecules in the room will congregate in one spot for an instant. This example is analogous to the probability that all information necessary to spawn a universe comes together at a single point, called a singularity. Oxford

Physics Professor Roger Penrose calculated the amount of information quanta in the singularity that made our universe. The number is a measure of the complexity of our universe. The number is one followed by the following number of zeros: a trillion times itself, ten times. This number is the largest in physics. The length of the number of zeros in the Penrose number, if it were written out, would be seven times the circumference of the universe. From this, it can be estimated that our universe was created by accident is less than one chance out of a Penrose number. This is the number behind the probability of the "lucky by accident" multiverse hypothesis. Roger Penrose said, "I think I would say that the universe has a purpose, it's not somehow just there by chance. Some people, I think, take the view that the universe is just there and it runs along, it's a bit like it just sort of computes, and we happen somehow by accident find ourselves in this thing. But I don't think that's a very fruitful or helpful way of looking at the universe. I think that there is something much deeper about it. Mathematical truth means more than just numbers. It reflects values of balance, harmony, logic, abstract beauty implies hope, love, compassion, consciousness. The force of love, hate, curiosity can't be measured by physical science. No matter how you look at it, the world is a mirage."

Avoiding Truth

"The fool says in his heart, there is no God." Psalm 14:1

In the 1980 TV series, cosmos, astronomer and atheist Carl Sagan told a rapt audience, "The cosmos is all that is, or ever was, or ever will be," concluding there was no need for a Creator; he was dead wrong. Edwin Hubble had discovered the big bang 50 years earlier. Now, Sagan's protégé, Neil deGrasse Tyson, has sheepishly admitted he is an atheist as well. He shared this metaphor with TV viewers: The unknown is like an ocean in which scientists have scooped and analyzed a cup of water to conclude whales are a myth. How can he not acknowledge God is the whale in his metaphor?

In terms of understanding our universe, proof positive of the big bang in 1965 changes everything, just as the Copernican revolution did five centuries earlier. Now physicists must think outside the cosmic bubble.

Logic requires what we think of as existence must still be infinite. For this reason, the idea of an eternal Creator of a finite universe, as revealed in Scripture, is more compelling than ever. But the God Hypothesis remains a threat to science; especially now there is far more reason to believe in God than not.

CNN: 9/2/2010 God did not create the universe; world-famous physicist Stephen Hawking argues in a new book that aims to banish a divine Creator from physics. Hawking says in his book, "Grand Design," given the existence of gravity, "the universe can and will create itself from nothing," according to an excerpt published in The Times of London. "Spontaneous creation is the reason why there is something rather than nothing, why the universe exists, why we exist," he writes in the excerpt. "It is not necessary to invoke God to light the blue touch paper (a British fuse) and set the universe going," he writes.

According to Hawking, the design of the universe is not "grand" enough to deserve a designer. So, let's take a look at his theory compared to the Bible. In physics, the word "nothing" is too ambiguous to have any meaning. Even if you have nothing in your wallet, you still have air in it. Even the vacuum of space has something in it. No place in our universe has nothing in it. If the explanation for existence is going to start from nothing, we must begin from non-existence, but non-existence has no physical properties that physicists can observe, so non-existence is meaningless in physics.

The Bible tells us God created existence. The word existence as used here means the existence of something of a physical nature. The Bible makes it very clear God himself does not have a physical nature. The Bible tells us God is a mysterious eternal Mind similar to our minds. Mind is not material; the mind is the symbolic content of the brain. If we are going to assume there is no God, another explanation for why anything exists is needed. Physics fails to have any such explanation. For Hawking to assume gravity always existed is a gross violation of the principles of science. The Bible states God began creation starting from a void. The void is a right word for something that exists but has no physical properties. So even a void has no meaning in physics because it has no observable properties.

Already, the God Hypothesis offers answers that cannot be found in physics. Physics has discovered laws govern the universe, but physics cannot tell us where these laws came from. The Bible tells us God is the great Lawgiver; the world was created from nothing but God's spoken word. Without a shred of evidence, Hawking's theory, now peddled as good science by other physicists, assumes an infinite gravitational field full of universes that must have existed before our world could be created. Then the magical "fuse" thing happens. Extrapolating one creation event into an infinite number of creation data points is not credible science. And even worse, nothing is anything; it is a whole lot of something.

Philosophers Leibniz and Sartre said the most significant philosophical question is, "Why is there something rather than nothing?" More something as proposed by Hawking and other prominent physicists does not solve the mystery of why there is something rather than nothing.

There will never be an experiment to prove Hawking's theory. The test that demonstrates the truth of the Bible is called life. We each get one life to decide God is a truth or a lie. There are no other possibilities. Who would you trust with your life, Hawking or the God of the Bible? Mark Twain said, "The two most important days in your life are the day you were born and the day you find out why."

Sean Carroll's recent book, "The Big Picture," [39] is a thinly disguised attempt to exclude God from the big picture. The Oracle of atheism, Christopher Hitchens, prominently displayed his prejudicial bias by claiming brilliant French physicist Pierre Laplace was an atheist, just because he told Napoleon he did not need God to explain planetary motion. Every prominent physicist from Copernicus to Einstein publicly declared their belief in God. Einstein said, "Science without religion is lame; religion without science is blind." [6]

Atheist and physicist Lawrence Krauss crudely mock Christianity. He claims science is the only source of real knowledge. He openly admits the purpose of his research project is to prove there is no God.

Krauss wrote a 2012 bestseller, "A universe from Nothing." [36] Since the universe is mostly nothing, why don't we see universes being created everywhere like popcorn? Perhaps we can get things started by shooting some equations into outer space. His title should have been "Something from anything but God."

Causality means everything that happens must have a cause. An accident has a reason which is unknown or yet to be determined. Causality is a fundamental principle of physics, but causality breaks down when explaining the primary or first cause, which can have no physical reason by definition. Thus, the Uncaused Cause cannot be included in the physics of any causal chain. Therefore, any scientific theory of origins must consist of the God Hypothesis as the Uncaused Cause.

Renowned Scottish physicist James Maxwell discovered the laws of electromagnetism in the late nineteenth century. He had this to say, "Science is incompetent to reason upon the creation of matter itself out of nothing. We have reached the utmost limit of our thinking faculties when we have admitted, because matter cannot be eternal and self-existent, it must have been created. I have looked into most philosophical systems, and I have seen none will work without God."

British biologist Richard Dawkins is the self-appointed high priest of atheism who preaches his beliefs with the enthusiasm of an evangelist. He has created a frenzy over his commitment to what he believes, which is evolution proves there is no God. And it seems most biologists agree with him. Evolution is a scientific fact of profound importance. The biological secrets of evolution stand ready to revolutionize medicine. God has given science the keys to the cosmos. Let's hope and pray we use them wisely.

In his book, "The God Delusion," [40] Dawkins claims Einstein was an atheist who only used the term god as a metaphor for nature. Einstein told Prince Hubertus, "In view of such harmony in the cosmos, which I, with my limited human mind, am able to recognize, there are yet people who say there is no God. But what really makes me angry is they quote me for the support of such views." [4] He went on to say, "The militant atheists are like slaves who are still feeling the weight of their chains, which they have thrown off after a hard struggle. They are creatures who, in their grudge against the traditional opinion of the people, cannot hear the music of the spheres."

Science will never tell us why anything exists, let alone the meaning of existence. But if you believe in God, you think He is why we are here, and you know He thinks in science because He designed and invented the cosmos. Science proves the universe is intelligible and therefore the world was conceived by an Intelligent Mind. Scientists are not the inventors of the universe; they are just forensic investigators looking for clues about how nature works.

The cosmos is a magical mystery theater for humanity to explore. It is so big we will never tire of searching for its secrets. The earth is just a place for humans to live while they figure out why living is important. We have only one life to get it right. Destiny is waiting. Wormfood does not have to be the end. Caterpillars don't know they get to be butterflies.

Origins Project

"Forget Jesus. The stars died so that you could be here today." Lawrence Krauss

Many physicists have joined the No-God movement. The tactics are new but the strategy is two centuries old. Their mantra is that only science is a source of real knowledge; they believe biblical revelation is fiction.

Let me introduce the self-anointed evangelists of the new atheism. They call themselves the "Four horsemen." Richard Dawkins, a biologist, is the high priest. His disciples are Sam Harris, a neurologist, Daniel Dennett, a philosopher and Christopher Hitchens, a polemicist. Harris condemns Christianity as a scourge on society. He wrote, "If anyone has written a book more critical of religious faith than I have, I'm not aware of it. Faith is belief without evidence." I must report that faith is belief with scientific evidence. Hitchens died of cancer in 2011 at age 62. His mantle was bestowed on Lawrence Krauss. Here is an excerpt from his bio on Wikipedia:

Lawrence Krauss is an American theoretical physicist and cosmologist who is Professor of the School of Earth and Space Exploration at Arizona State University and director of its Origins Project. He is known as an advocate of the public understanding of science, of public policy based on sound empirical data, of scientific skepticism, and scientific education, and works to reduce the impact of superstition and religious dogma in popular culture. Krauss is an atheist activist and self-described antitheist. Krauss has participated in many debates with theologians and apologists.

His book, "A Universe from Nothing: Why there is something rather than nothing," [36] with an afterword by Richard Dawkins, was released in January 2012 and became a New York Times bestseller within a week. Dawkins claims the book is comparable to Darwin's.

Because "it upends the last trump card of the theologian."

Consider the assertion, "if nothing then universe," made by Krauss in the title of his book. In order to validate this hypothesis, it must pass the falsification test as follows: "If nothing then no universe." There is no empirical way to test for "no universe." The falsification test fails. Recall the black swan test. The proposition, "All swans are white' contains meaningful content because the proposition, "All swans are not white" is also meaningful. Now consider the proposition "If nothing then no universe." This proposition contains no empirical content because "nothing" has no properties, therefore cannot be observed. The term "no universe" has the same problem. the black swan test reveals the logical and scientific absurdity of the proposition "The universe was created from nothing." The God Hypothesis does not have this trap. Because God and existence are infinite and our universe is finite, God created the universe and the not-universe is a logical and meaningful claim. QED

Now Krauss claims the "nothing" of which he speaks is "relativistic quantum fields." Calling this "nothing" is false prima facie. "Nothing" in this context means nothing, as in non-existence. Krauss is using sleight of hand to pedal atheism. He concludes, "I don't ever claim to resolve that infinite regress of why-why-why-why-why; as far as I'm concerned it's turtles all the way down." In a 2007 New York Times article, Krauss put it this way: "We're just a bit of pollution." The original physician without borders, Albert Schweitzer wrote, "A man is ethical only when life, as such, is sacred to him." Instead of Origin, Krauss found out how to go to hell.

Physicist David Albert reviewed Knauss' book: "It has nothing whatsoever to say on the subject of where those fields came from, or of why the world should have consisted of the particular kinds of fields it does, or of why it should have consisted of fields at all, or of why there should have been a world in the first place. Period. Case closed. End of story."

Richard Dawkins cites Krauss by summing up the mindless accidents hypothesis this way; "The universe we observe has precisely the properties we should expect if there is, at bottom, no design, no purpose, no evil, no good, nothing but blind, pitiless indifference." Biologists have had their day defending evolution against creationism. This debate was over a century ago. Creationism is simply bad science. But the fact evolution won is a confidence builder for atheists.

Richard Dawkins claims to disprove God in his book, "The Blind Watchmaker." [41] He points to flaws in the design of the human eye as evidence. He states, "the human eye is the result of mindless evolution, not intelligent design." Dawkins is the one who is blind to the truth. The author has AMD and glaucoma. Why doesn't Dawkins do something about flaws of the eye, instead of criticizing God, who he doesn't even believe in? The Bible is concerned with why God made humans, not how He did it. The Bible does not claim our creature nature is made in the "image of God." We are flawed creatures. It is our hearts and minds, which are made imago Dei. The duality of human nature takes us out of nature while we still must live in it. This is the human condition we all must overcome.

God's purpose cannot be found in the design of material objects such as eyeballs. What we can learn from this example is that Dawkins has been blessed with a good mind. By the end of this century, men like him will have the biological know how to improve the design of the human eye, although there surely will be more pressing issues. God has given us the power to solve any problem if we as a human race, have the will to do so. We have the power to make a better world or destroy it. We are the gods of our destiny in this universe. We must save ourselves while we are here. God's kingdom is not here.

According to Dawkins, "A good and just God would never allow havoc of suffering, waste, and death wreaked by evolution." This is a bogus argument. Evolution is a process that has little or no bearing on life in a civilized society. Evolution of our species is a relic of our pre-human past. Animals don't spend their time thinking about death like humans do. The suffering and death of humans at the hand of nature and humans strikes anguish and grief in the hearts of all decent people. But this has nothing to do with evolution. Evolution now serves as a magnificent gift; knowing how evolution works is a new and emerging benefit to mankind. We are eternal beings caught in a mysterious epic battle between good and evil.

We can be saved only by the grace of God. The agonies of life draw us together as we seek mercy from God. There are real reasons why we should believe there is a just and more perfect world beyond the grave to give us any hope to overcome the heartaches, misery, and suffering we face in life. Dawkins should heed the words of Bohr to Einstein, "Who are you to tell God what to do."

Here is the dogma of New atheism:

(1) Reduce reality to mindless atoms,

(2) Reduce creation to a mindless accident,

(3) Reduce the value and meaning of human life,

(4) Denigrate the value and purpose of religious faith,

(5) Denigrate attempts to find meaning in science.

Institute for Creation Research

"It isn't what we don't know that gives us trouble; it's what we do know that ain't so." Will Rogers

7/26/1925, Dayton, Tenn. John Scopes was found guilty of teaching evolution in a public classroom. "I think this case will be remembered because it is the first case of this sort since we stopped trying people for witchcraft because here we have done our best to turn back the tide that has sought to force itself upon this–upon this modern world, of testing every fact in science by a religious dictum." Clarence Darrow, John Scopes' attorney, added an interesting side note to this trial occurred when Darrow went for the jugular. He asked the prosecution's famous attorney, William Jennings Bryan, to explain how snakes walked before God cursed them and made them crawl on their bellies. Bryan said he did not know the answer. Darrow then asked Bryan if perhaps they walked upright on their tails. The question caused a huge outburst of laughter from the courtroom. Interestingly, we now know primeval snakes walked on four legs because they have vestigial leg bones on their bellies to prove it. For a person to say he doesn't believe in evolution is tantamount to saying he does not believe in biology or any other science for that matter.

The Institute for Creation Research began in San Diego in 1970. It soon moved to Dallas, Texas, where they found a more receptive climate. Their core belief is the Bible is literal truth from God that trumps all human knowledge. The ICR charter involves lectures and literature targeting the evangelical community and religious schools with perverted scientific propaganda. ICR operates a $30 million Museum of Creation and Earth History, illustrating the six-days of creation that occurred 6,000 years ago. They also put on biblical adventure tours. The faithful contribute over $8 million a year.

It should come as no surprise the author finds ICR dangerous and highly offensive. They presented a movie and seminar at my church called "The Truth Project." The purpose was to convince the audience evolution is a lie. My family members subscribe to ICR's monthly publication, "Acts & Facts." My parent's' minister told them I put science above God. This was enough reason to write a book. The author felt his book would not be fair unless it was an equal opportunity offender. In the 1987 case of Edwards v Aguilars, the U.S. Supreme Court ruled creationism is religion, not science, and teaching it in public schools is unconstitutional!

In the book, "Denialism", Michael Specter [42] discusses in great detail how irrational thinking hinders scientific progress, harms the planet and threatens our lives. With tongue in cheek, he writes the Christian right would go so far as to denounce electricity as a force of evil, blamed its prevalence on liberal socialistic utility companies, universalizing the relatively rare horrific experiences of people who have been injured by electrical currents and call for a ban on electricity use. Mr. Specter defines denialism, as what happens when an entire segment of society, often struggling with the trauma of change, turns away from reality in favor of a more comfortable life.

Reality Check

"By faith, we understand that the worlds were framed by the word of God, so that the things, which are seen were not made of things, which are visible." Hebrews 11:3

"When an astronomer writes about God, his colleagues may assume he is either going over the hill or going bonkers. In my case, it should be understood I am an agnostic in religious matters. But I would like to pursue scientific inquiry further by going back in time to the beginning, to the actual creation event. But we cannot see it because the massive energy field of the big bang is plasma composed of a soup of pure energy and charged subatomic particles, which are opaque. The barrier to further progress seems insurmountable. It is not a matter of another year, another experiment, another measurement, or another theory. At the moment, it appears scientists will never raise the curtain on the mystery of creation.

For the scientist who has lived by faith in the power of reason, this is like a bad dream. He scaled the mountain of ignorance; he is about to conquer the highest peak. As he pulls himself over the rock at the pinnacle, he is greeted by a band of theologians who have been sitting therefore centuries."

Robert Zastrow, Director of NASA Institute of Space Studies

Atheists believe only science and reason are sources of knowledge and all of human experience can be reduced to the material world. For over two centuries the false religion of reductive materialism has enshrined atomic particles as the accidental source of the universe. Reductive materialism claims that there is no independent, autonomous level of phenomena in the world that would correspond to the level of conscious mental states. It claims the level of conscious phenomena is identical to some level of purely neurological origin. Conscious phenomena are nothing more than neurons firing. This theory seems to leave out what it is like to experience something that cannot be described like the smell of a rose.

Even you can't be sure what you really think is reality. Is all that exists described by the reality we observe? We are not mind readers, at least not yet. Who can we trust? Science is complementary to the biblical view of the world. What you see and think are your own reality. The notion of discrete matter dates back to ancient Greece. Even today, many scientists and philosophers explain reality in terms of discrete particles. This idea, supported by classical physics, requires the addition of invisible forces to the particles to connect them together. But the idea that atoms are like little billiard balls has been discarded by the discoveries of the last century. Particles are no longer isolated entities. The universe can no longer be reduced to independent atoms. Gravity and quantum physics connect every quantum field in the universe to every other quantum field in a unified whole. In his 1932 Nobel laureate speech, Werner Heisenberg said, "The atom has no immediate or direct physical properties at all." To quote Neil's Bohr, "Isolated material particles are abstractions, their properties being definable and observable only in terms of their interactions with other systems." Science writer David Bradley put it this way, "We may agree, perhaps, to understand by metaphysics an attempt to know reality as against mere appearance, or the study of first principles or ultimate truths, or again, the effort to comprehend the universe, not simply piecemeal or by fragments, but somehow as a whole."

What we see and call particles is not what is actually there. We only see light waves interacting with quantum fields, which makes them appear as matter. Quantum fields create energy patterns that appear as matter as per e = mc2. Quantum fields are the self-supporting medium of quantum particles. The web of interconnections even includes the observer. The observer is now an integral part of scientific experiments. His mind is literally involved in deciding the outcome of an experiment. Instrumentation cannot replace the conscious mind of the observer. As stated by Einstein; "Since the theory of general relativity implies the representation of physical reality by a continuous field, the concept of particles or material points cannot play a fundamental part and can only appear as a limited region in space where the field strength and energy density are particularly high."

Physical science is converging with philosophy. The scientific understanding of reality is not opinion but scientific fact. Occam's razor brings science into harmony with metaphysics and the foundation of all knowledge. There must be one substance that causes and connects the totality of what we experience. As we all experience many minds and many material things but always in one universal cosmic spacetime, we must understand reality in terms of the one common spacetime of cosmic quantum waves we all share. The wave structure of cosmic reality is the central theme of relativity, cosmology, and quantum physics.

Since prehistoric time, human intellect, self-awareness, and the search for meaning have separated us from all other creatures. Humans have always pondered a greater reality that went beyond the five physical senses into a mystical sixth sense, a mind, which deeply unites our sense of existence with an unseen reality hidden within physical reality. All primitive cultures were in awe and fear of the unknown, which seemed to be ever present. Both Eastern and Western cultures have developed profoundly mysterious and transcendent views of reality, which have little or no relevance to the more mundane mechanisms of biological evolution. The universe has been evolving for 13.8 billion years. The biological evolution of humans would only be the last minute on a 24-hour clock.

The unavoidable conclusion is that science has discovered the hidden nature of reality, which can only be attributable to the Creator, as mankind has believed in one form or another since time immemorial. Science has literally discovered the reality behind philosophy, religion, mysticism, superstition and wishful thinking. Science has reached the limits of material reality and cannot penetrate the mystery of reality just beyond its grasp. Science is poised to expand its scope of study beyond physical reality into the supernatural world of non-physical forces. Our existence in the universe is nothing less than amazing. We are subtly connected to all things around us. This is fundamental to why we can see and interact with our world.

The central problem of metaphysics has been solved. The one active substance that causes and connects the many changing material things we experience is simply the source of all physical reality, the information of the cosmic code. The observable universe exists as a finite region within an unbounded, eternal reality. God is the source of the eternal reality and the mind of God is the creative essence behind all of reality. God's will has conceived of the process of evolution as the mechanism to form the universe with creatures, imago Dei, as its purpose. Let us briefly consider the role of biological evolution in the creation process.

8/10/2009 USA Today, "Nearly half of Americans still dispute the indisputable: that humans evolved to our current form over millions of years. We're scientists and Christians. Our message to the faithful: fear not." Professor Darrel Falk, Point Loma Nazarene University, San Diego.

Atheists like Richard Dawkins claim evolution is the result of pure chance, which "proves" there is no God. In actuality, evolution is not pure luck. From ancient times to the present, Plato's teleological argument asserts we live in a purpose-driven universe.

In evolution, a single mutation can change the course of life. Therefore, God has the means to intervene in nature at any time without interfering with any of the laws of nature. God can leave the little details of nature to the original self-fulfilling design. Thus, divine purpose can be fulfilled with or without direct intervention. God's grand design allows maximum, moral freedom and openness. Moral norms may change, but this implies humans trust their better nature. Werner Heisenberg said, "Natural science is the foundation of technically appropriate action; religion is the foundation of ethics, as opposed to psychoanalysts who attach religion to obsessive disorders, neuroses, infantile fixation, repression, and self-deception. Belief in God gives meaning to life and a will to do good." Theologian and astronomer William Steger weighed in with this thought, "God acts through the laws of nature. If we put this in an evolutionary context, we can conceive of God's continuing creative action as being through the natural unfolding of nature's potentialities." Physicist Ian Barbour had this to say, "What appears to be chance, which atheists take as an argument against theism, may be the very point at which God acts. So, God's actions would be scientifically undetectable. It could neither be proven nor refuted by science." [43] In actual fact, evolution is just a continuation of the self-organizing principles, which created the entire universe. The ultimate question is, from where did those self-organizing principles come? Genesis asserts God created order out of chaos. Atheism-of-the-infinite -gap claims the universe created itself from nothing. Only one can be right. Which makes the most sense to you?

It is obvious that advances in civilization are the end result of an intelligent knowledge accumulation process. It just so happens the progress of human knowledge takes place by self-organizing principles analogous to those that guide the natural development of our place in the universe. Plato said, "necessity is the mother of invention," but Thomas Edison proved trial-and-error is the method of invention and discovery. Trial and error are the most powerful strategy for learning and progress known to mankind because of the cumulative effects of gaining knowledge acquired incrementally over countless generations. We know the goals of civilization and the challenge is to continually strive for progress towards achieving those goals. It is the same way each person guides his or her life. Making mistakes is fundamental to the learning process in nature as well as in our lives. Parents are proud to see their children learn from their mistakes. So, it is with our Heavenly Father. God created the universe with a goal and the universe is fulfilling that goal by producing conscious, intelligent beings. God knew exactly what He was doing in spite of what Dawkins thinks.

CHAPTER 7

COSMIC MIND

Way Forward

It used to be that scientists believed everything was simply constructed from other smaller things, ultimately atomic particles. Everything behaved in a certain complex way because its smaller parts behaved in simpler ways. Thus, a human organ functions the way it does because its molecules behave the way they do. And molecules mindlessly behave themselves because they are composed of equally mindless atoms. Atoms are fundamental units of lifeless material substance mindlessly obeying the mechanistic laws of the universe. Ultimately, all things in the whole dead universe behave mindlessly, following the mechanical laws that Newton so carefully explained over three centuries ago.

Of course, biological systems are special. They are alive. Yet we ultimately explain living systems as emerging from patterns made by dead mechanical pathways that their simpler molecules follow. Of course, mapping these pathways would be a monumental task, perhaps impractical or even impossible. But today, with the rapid advances in knowledge of DNA and biomolecular science, and with the amazing advances in computer technology, one might begin to think that such an undertaking is possible. Even if it is, however, no one has yet seen an equation get up and walk. The origin of life and consciousness are still a mystery. And materialism has had a death blow. The simplest units of existence, subatomic particles, and atoms made from them, could care less about the ideas of Democritus or Newton. Instead, atomic particles capriciously follow a myriad of paths of their own choosing as they move from place to place. Not only that, but somehow the mere act of observing such particles upsets these paths and alters their history. Even more mysterious, atomic particles behave like ethereal quantum waves until a human mind observes them under close inspection, and thereby, make them appear to behave like particles. Schrödinger wrote, "The verbal interpretation, i.e. the metaphysics of quantum physics, is on far less solid ground. In fact, in more than 40 years, physicists have not been able to provide a clear metaphysical model."

Sometimes called the principle of indeterminism, the uncertainty principle reflects the inability to predict the future based on the past or the present. Known as the cornerstone of quantum physics, it provides an understanding of why the world appears to be made of events that cannot always be connected in terms of causality. In essence, it says, let uncertainty stand between us and the true nature of reality. The complementarity principle states the physical universe can never be known independently of the observer's decision of what he chooses to observe. These choices fall into two distinct, or complementary categories of observation. Observation and determination carried out using one category always precludes the possibility of simultaneously observing and determining the complementary category. For example, the position of an object and the path it follows through space and time are observations in complementary categories and so cannot be determined simultaneously.

The following is a visual analogy to a complementarity quantum process within the mind. This may indeed reflect a real quantum physical process in your mind.

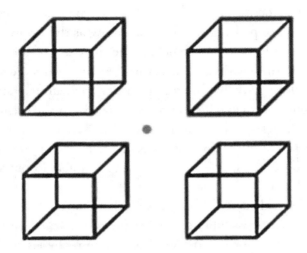

Necker Cubes

Look at the cubes. Which face of the cubes is in front? Are you looking down on the cubes or looking up at them? You should see them looking down from above because the mind expects the cubes to be sitting on something. You can only see the cubes facing in unison. How you see the cubes depends on how you choose to see one cube. You can't see both ways at the same time. You either see cubes facing up or facing down. They take on a duality of entangled states in your mind.

According to the observer effect, the cubes only become a real cube, with one side forward, when you observe them. In this interpretation, you cause the event to occur, such as seeing a specific view of the cubes. Bohr said, an act of observation involves a merger of two ways of seeing: the classical world of perception and the atomic world of quantum events. By attributing the effect of an observer to an intrinsic eventual merger of the quantum world with the classical world, the question of the nature of the act of observation is left open. Bohr found when an observation takes place, the property of the object under scrutiny mysteriously takes on a specific value. This is to say, we don't know how an act of observation takes place. All we know, is it is ultimately an unconscious decision of the mind acting outside the mind on an object. You will also notice you are looking at lines on paper, but your mind can only see 3D objects because this is what it expects.

Visionary physicist John Wheeler explores what quantum physics and information theory have in common regarding the age-old question, "How come existence?" He concludes the world cannot just be a giant machine ruled by laws. The laws of science, as we know them, provide mere approximations to the true information source from which they derive. Fundamentally, there is no physical universe without the elementary act of observer participation. John Wheeler's interpretation epitomizes this in biblical terms. He claims our words shape and change reality. Each of us must take responsibility for our own words and actions. It is like we were made imago Dei. It's as if we each make our world the way it seems we want it to be in our minds. And the way it seems springs from our actions, which are, in turn, conditioned by words and symbols we use to make our world what it seems to be. Even so, why do we choose to do the things we do? Does God have anything to do or say about the actions we take? It seems apparent some gentle force is guiding our hearts and minds. How does this force come into being? How can we discover it?

Necessary Being

"How can it be demonstrated that God, without any change in himself, produces effects subject to change and measured by time?" Saint Augustine

It makes logical sense that in order for anything to exist, something must exist by necessity. There is nothing that has been uncovered by science that is more necessary than God. Science requires us to believe that the cosmos or any other possible physical entity must have a cause. Therefore, they do not exist by necessity because their existence is contingent on their cause. This cosmological argument for God is contained in the works of Plato, Aristotle, Aquinas, and Leibniz. The theory of necessary existence is rational, but it cannot be tested or confirmed by empirical methods. It goes like this:

- Every finite entity is contingent on its cause.
- A causal sequence cannot be infinite.
- Therefore, a first cause is necessary.

According to the argument, the existence of the cosmos or any physical entity requires an explanation, and the creation of the universe by a first cause is generally assumed to be God. In this context, a relevant description of God is as follows; God is necessary, immaterial, immutable, eternal, omnipotent, omnipresent, and omniscient. Recall God has the same attributes as his creation, except God is necessary, immaterial, and eternal. He has other qualities that define his relationship to man, such as, we are made imago Dei. In any case, it cannot be said that the God of Abraham is not believable because He is too vague, as Paul Tillich claimed.

If we are to believe God is the first cause (i.e., uncaused Cause), then "Who made God?" This age-old riddle threatens to throw reason into an infinite regress. The only way out is to assume that God can explain himself. This means we can only know God by what He has said and done. This does not mean our finite little minds can comprehend God. Saying that God is self-explanatory is almost self-evident because similar concepts arose in all early cultures. The Holy Scripture happens to be the most credible self-expression of God. With the sophisticated knowledge of today, we have the extraordinary privilege of examining the truths or myths of Scripture through the lens of science. Even without considering scientific evidence, logic, and reason require that it be established that only God can be the one necessary being. If God were not necessary, then the chain of causality could not begin or be attributed to God, and we would want to know what was beyond God, which explains his existence. The concept of a necessary being that always existed or somehow caused his existence has confounded theologians and philosophers since time immemorial. It is certainly an idea, which seems to be outside the bounds of logical reasoning. But the fact is, there is no other credible explanation for existence. Our minds are so intricately bound with time it is difficult to conceive of an existence unbounded by time. Gottfried Leibniz wrote, "all contingent things, even if they have existed eternally, lack adequate explanation until they are grounded in something necessary." This is called the argument of sufficient reason. J. L. Mackie argues, it is reasonable for something contingent to exist without reason for its existence. This argument defies logic because the term contingent existence means dependence on something else for existence.

Religious philosopher Hugo Meynell states, "God, by his nature, is the Being whose understanding and will explain how it is and that it is of everything else without himself being capable of being explained in the same way. That on whose existence, understanding, and will everything else depended could not be dependent on the existence of anything else." Atheist Hugo Parsons points out, "There is nothing in science that entails everything must eventually be explained." This is the mantra of atheists who should know better than ever to expect science ever to find a replacement for God. Parsons does not understand that the highest aspiration of science is to know the mind of God, as expressed so well by Einstein, "I want to know how God created this world. I want to know his thoughts; the rest are details." [4]

The unique contribution of monotheistic theology is the creation doctrine derived from Genesis. God spoke the cosmos into existence ex nihilo. Not only does ex nihilo agree with the latest theory of inflationary cosmology, but it also brings creation into existence, outside of time, a material universe created by divine power as an act of free will. By definition, free will involves chance because we say that a choice is free only if it could have been otherwise. So, if God is endowed with the freedom to choose between possible alternative universes, the contingency of the actual universe is explained. This means the universe does not have to have a particular order. Since the order of the universe does not exist out of necessity, it cannot be deduced by a priori logic. No logical proof can explain existence. Instead, it must be discovered using a posteriori reasoning. The laws of nature create and preserve an orderly and intelligible universe. This fact demonstrates that God makes rational decisions. These same laws allow openness by invoking natural evolutionary processes, ensuring random variability and diversity in all of nature. Cosmic purpose does not need to follow a precisely predetermined plan. "We are all unique. Even the eyes on the same person are different. I never paint them the same." Pablo Picasso

Creatio ex nihilo almost solves the paradox of how a timeless, necessary Being can explain a changing, unnecessary world. If the nature of God is necessary and unique, how could He have chosen to create a different universe than the one we have? Einstein raised this question. God must also have free will to create any universe He wants. If his choice was not rational but whimsical, we might as well be content with an arbitrary universe and leave it at that. British philosopher and theologian, Keith Ward, has studied the issue of God's necessity and the contingency of the universe. He summarizes the essential dilemma as follows:

"First of all, if God is self-sufficient, like the axiom of intelligibility seems to require him to be, how can it come about that He creates a world at all? It seems an arbitrary and pointless exercise. On the other hand, if God is a necessary and immutable Being, how can He have a free choice, surely all that He does will have to be done of necessity and without any possibility of alteration? The old dilemma, either God's acts are necessary and therefore not free (could not be otherwise), or they are free and therefore arbitrary (nothing determines what they shall be), has been sufficient to impale the vast majority of Christian philosophers down the ages. The problem is, whichever way you cut it, you come back to the same basic difficulty, that the truly contingent cannot arise from the wholly necessary.

"This view is in tension with a central strand of the Christian tradition: namely, that God need not have created any universe and that He need not have created precisely this universe. How can a necessary being be free in this way? Help came, however, from an unexpected source, quantum physics. The central message of quantum physics is Heisenberg's uncertainty principle. For a general quantum state, it is impossible to say in advance what value will be obtained by measurement; only probabilities can be assigned. Thus, when one measures a state, a range of outcomes is available. The system is therefore indeterminate, one might say, free to choose among a range of possibilities and the actual outcome is contingent. On the other hand, the experimenter determines what is to be measured, so an external agent decides the class of alternatives. As far as the particle is concerned, the nature of the alternatives is fixed by necessarily, whereas the actual alternative adopted is contingent. Suppose we ask a question of nature and get an answer. The answer we get is determined by the question we ask. Likewise, if we ask nature a series of questions, the final answer will now be determined by the series of questions asked. Quantum measurements reflect this reality. The final state of a quantum particle is predictable only to within a set of values determined by the sequence of measurements made by an observer in an experiment.

"Let us now return to the theological analogy. This mixture of contingency and necessity corresponds to a God who necessarily determines what alternative worlds are available in nature, but who leaves open the freedom of nature to choose from among the alternatives. This is called process theology. In process theology, the assumption is made that the alternatives are necessarily fixed to achieve a valued result (i.e., they direct or encourage the otherwise unconstrained) universe to evolve toward something good.) Yet within this directed framework, there remains openness. The world is therefore neither wholly determined nor arbitrary, but an intimate combination of chance and choice." [44] This methodology allows for diversity within the constraints of an intended goal.

So far, this cosmological argument proves that God's existence, as established by monotheistic theology, is not inconsistent with the scientific understanding of the universe. It even explains why evolution is an imperfect process yet results in fulfilling God's purpose. However, this argument does not establish that God's existence is a logical necessity. Scientific methods cannot logically prove or disprove the existence of God. Science also cannot logically prove the existence of existence itself. As discussed in the section on logical reasoning, we can deduce an effect only if we know its cause. But we cannot understand the cause from the effect. We can use physics to infer the cause of the big bang, but the cause cannot logically deduce the cause itself. Therefore, we can infer, but cannot prove, how the cosmos came into existence. If God is assumed to be a possible first cause, his existence cannot be confirmed, but it can be inferred from the totality of evidence available to us. No other hypothesis fits the facts. So, the fact is God can be known through the study of his handiwork. God sits at the apex of existence, and his word is the bedrock that sustains existence by his will and authority.

"For since the creation of the world God's invisible qualities, his eternal power, and divine nature, have been seen, being understood from what has been made so that people are without excuse." Romans 1:20.

Chosen Species

"It is hard to believe our reasoning power was brought by Darwin's process of natural selection, to the perfection which it seems to possess." Eugene Wigner, Professor of Physics.

The study of nature is interesting, but it is just the stage on which the real drama is played. You see, we are not like other animals. Humanity is the intended beneficiary of the emergent cosmic complexity. God chose the elegant design strategy of evolution to prepare the way for the arrival of his chosen species.

Our evolutionary heritage prepared us in every way to aspire to be one with nature. All of life on earth, from pond scum, to primates, are our family. Evolution gave us our beautiful mind and body, our biology and chemistry, our survival instincts and social skills. But we are not God's robots. We were prepared by our Creator to bestow on us, free will, knowledge of right and wrong, good and evil, an immortal soul, and the opportunity to be included in God's family. "God created man in his image, in the image of God He created him; male and female He created them." Genesis 1:27. The Torah is written without vowels, which tends to make its true meaning more mysterious and wondrous. It makes no difference whether the creation stories are literal or symbolic. Either way, the meaning of our place in creation is clear enough. In the words of C. S. Lewis:

"For long centuries, God perfected the animal form, which was to become the vehicle of humanity and the image of Himself. He gave it hands whose thumb could be applied to each of the fingers, and jaws and teeth and throat capable of articulation, and a brain sufficiently complex to execute all the material motions whereby rational thought is incarnated. The creature may have existed in this state for ages before it became man; it may even have been clever enough to make things, which a modern archaeologist would accept as proof of its humanity. But it was only an animal because all its physical and psychological processes were directed to purely material and natural ends. Then in the fullness of time, God caused to descend upon this organism, both on its psychology and physiology, a new kind of consciousness, which could say I and me, which could look upon itself as an object, which knew God, which could make judgments of truth, beauty, and goodness, and, which was so far above time, it could perceive time flowing past. We do not know how many of these creatures God made, nor how long they continued in the Paradisal state. But sooner or later, they fell. Someone or something whispered they could become as gods. They wanted some corner in the universe of which they could say to God; This is our business, not yours. But there is no such corner. They wanted to be, nouns, but they were, and eternally must be mere adjectives. We have no idea in what particular act, or series of acts, the self-contradictory, impossible wish found expression. For all I can see, it might have concerned the literal eating of a fruit, but the question is of no consequence." [5] C. S. Lewis.

The story of Adam and Eve is a study of the nature of humans. The meaning of human plight is laid bare in a simple tale of high moral drama. Every person is responsible for his actions and ultimately accountable to God. Every person relives the story of the fall from grace, rejection, despair, and the need for redemption.

Science has unveiled the cosmic events leading to the fulfillment of God's foreordained plan. Consider the following passage from the Bible: "Springs came up from the earth and watered the whole surface of the ground—the Lord God formed the man from the dust of the ground and breathed into his nostrils the breath of life and man became a living being." Genesis 2:6,7.

This amazing passage makes it clear God made all life, including man, from inanimate particles of dust. This Scripture foretells the notion of life conceived out of the primordial soup. Genesis 1:20 & 1:24 have a Hebrew phrase meaning "living being" This is used to describe the man as well as fish, birds, and animals. "The likelihood of the formation of life from inanimate matter is one to a number with 40,000 zeros after it. It is big enough to bury Darwin and the whole theory of evolution. If the beginnings of life were not random, they must, therefore, have been the product of purposeful intelligence." [34] Sir Fred Hoyle.

According to Genesis, Adam was the first spiritually aware human, but he was not the first member of his species. Adam is described as the predecessor of the Jewish people who lived near the dawn of civilization sometime around 4,000 BCE, according to biblical genealogy. There is nothing about evolution that contradicts specific claims in Genesis. If God created life through the process of evolution, we should embrace our dominion and oneness with all of life. The unity of life on earth reinforces the unity and harmony of all of creation.

No other creature has a mind capable of knowing God and a soul searching for meaning and purpose in life. We, as humans, have been given the opportunity to rise above our creature nature. We are here to prepare us to stand in God's presence. We have free will, the power to choose our destiny for eternity. God gave his chosen people a way forward to ensure his plan could be fulfilled.

According to the Bible, Adam and Eve had two sons, Cain and Abel. Out of jealousy, Cain killed Abel, at which point Cain fled Eden and married a stranger. "So, Cain went out from the Lord's presence and lived in the land of Nod, east of Eden. Cain lay with his wife, and she became pregnant and gave birth to Enoch." Genesis 4:16. Clearly, there were other homo sapiens living at the time who were brought into the family of humanity.

Fundamentalists claim Cain married one of his sisters. This is impossible because Jewish law prohibits incest and they would not start their holiest book with such a story. Some fundamentalist Christians think reconciling the creation story with science sends us down the slippery slope of liberal theology, denying miracles and fundamental truths of the Bible. The fact is mature Christians have been living on this slippery slope for centuries as they cope with the struggle between faith and reason.

Human Paradox

"At the very moment of vain-glorious thought, a qualm came over me, horrid nausea and the most dreadful shuddering. I looked down. I was once more Edward Hyde." "The Strange Case of Dr. Jekyll and Mr. Hyde," Robert Louis Stevenson.

We have evolved out of nature, yet we are still in it. We are confounded by our bifurcated dual nature, shooting for the stars, yet stuck in a dirtbag. We live knowing our fate. We wear out and die. Other animals are spared our depressing dilemma. Nightmares haunt our dreams. After a night of grappling with made up worlds, the sun rises to see the order God created for us. Yet we think nothing of it. By day, depression is always lurking. We have a great desire for joy, but it is just a momentary aberration.

The ancient Greek philosophers, Jewish prophets, and the great men of the Renaissance fully recognized our godlike status. But Kierkegaard exposed the Achilles heel of human existence. Kierkegaard found man's existential dilemma is laid bare on the expulsion of Adam and Eve from the Garden of Eden. The fall of humans is the crucial insight of all psychology. Humans must endure the insufferable duality of their magnificently conceived intellect and their unholy creature nature. Humans are infatuated with themselves, and God gave them what they wanted. Humans were given the knowledge and freedom to create their own world. They were no longer animals living by instincts.

Kierkegaard recognized humans respond to their dilemmas by creating a façade of respectability. But human character is a carefully honed lie, which hides the truth about himself. Eventually, every person will experience a feeling of hopelessness, which is the true nature of existence. Finally, we must accept our decay and death. The sincerest moments of life occur when we face the superficiality of finite existence. The selfish ego must die before a higher reality can begin. Humans must learn to transcend their creature nature. The reason is a mental block to awareness of God. Rabindranath Tagore, an Indian Bengali put it well, "A mind all logic is like a knife all blade. It makes the hand that uses it bleed." Faith in God is an act of humility, love, and trust, as the unconditional love of a child for its parents. Jesus told Nicodemus, "No one can see the kingdom of God unless they are born again." John 3:3. It is never too late to leave a misguided life behind and start a new life, which honors God.

The existential crisis occurs when a person realizes his life has no real purpose. Kierkegaard exposes the dichotomy between reason and meaning. Logic leads to pessimism and the despair of death. Lack of real meaning is the existential crisis of life. "For what will it profit a man if he gains the whole world, and loses his soul?" Mark 8:36.

Recall Kierkegaard's metaphor of the mind being likened to a two-story house. On the first-floor reason is the guide to life. But something is missing, which can only be found upstairs where reason cannot go. The upper floor of non-reason transcends the profane and mundane world of existence. Hostility, suffering, and despair lead to the stairway. Upstairs, a phase change transforms a person into a lightness of being. Faith presents a spiritual reality that transcends the material world.

Mind opens to the cosmic reality of God's love, which gives meaning, purpose, values, compassion, unity, and oneness of humanity, E Pluribus Unum.

Freud laid bare the creature nature of humans. His associate, Otto Rank probed deeper to lend scholarly credibility to Judeo-Christian theology, and Kierkegaard's work [45] Rank, concluded a person who ignores God is missing the real meaning of life. The most profound experience in life is to know the meaning and fulfillment of agape love. "You are here to enable the divine purpose of the universe to unfold. That is how important you are!" [46] Eckhart Tolle, Oxford professor, spiritual teacher.

The competition between God's love, agape, and human love, eros, is the crux of the Adam and Eve story. God gave Adam free will, and Adam chose Eve over God. "Agape makes the world go around; eros makes the ride worthwhile."

Guilt-ridden man cannot overcome his selfish creature nature. How can a man deny his nature to find self-worth? Without God, man must create his own worth.

Culture is the civilized way to deny our creature nature. For centuries, Western civilization lived by cultural beliefs of the Judeo-Christian tradition. God's purpose was manifest in all things. We were made lowly creatures to seek salvation and fulfill God's higher purpose. Christianity had answers for all our problems. The Age of Reason banished these answers as unworthy of our intellect. Enlightened men rejected religion as weak minded. The comforting solace of God's pure love was gone. Instead, men found gratification in the love of another person.

Nietzsche despised Judeo-Christian morality. So did Bertrand Russell. Many atheists rejected religion because it constrained their evolutionary right of power, domination, and self- gratification. Sexual morality was against our nature, even though monogamy was needed to protect the family.

Freud believed surrendering to God was demeaning. Freud concluded the sexual conflict of the Oedipus complex explains man's ego dilemma. Modern psychology interprets the Oedipus complex more generally as a search for meaning in life. We must conclude it is demeaning to view humans as mere creatures of evolution.

Modern Man

Modern man trusts only his analytic mind. He rejects naïve hope as wishful thinking. He has a firm grasp on reality. He has banished mystery about life. He is confident in the certainty of his worldview. He has no doubts. Nothing can sway him. He knows he is an animal whose life serves no real purpose. There is no God. We are worthless scum. We might as well not have been born. We are an accident of nature. Our mind is nothing more than a neurological fabrication of evolution. The winner dies with the most toys.

Psychologists refer to this as a form of neurosis. It is an inability to change or adapt one's life patterns to a more vibrant, more complex, more satisfying life. The psychoanalytic theory considers it an ego defense mechanism. Eminent psychoanalyst Carl Jung finds this explanation particularly fitting for people who are successfully adjusted by usual social standards, but who nevertheless have issues with the meaning of their lives. "I have frequently seen people become neurotic when they content themselves with inadequate or wrong answers to the questions of life. The majority of my patients consisted not of believers, but of those who had lost their faith. Modern man is blind to the fact, with all his rationality and efficiency, powers that possess him are beyond his control. His gods and demons have not disappeared at all; they have merely got new names. They keep him on the run with restlessness, vague apprehensions, psychological complications, addictions, a vast array of neuroses." Carl Jung sees collective neuroses in modern society: Our world is, so to speak, dissociated like a neurotic.

The English writer and philosopher Gilbert Chesterton brought the ideas of Kierkegaard into modern psychology. He recognized modern man prides himself on logic and objectivity. Modern man is serious by nature. He does not trust anything which appears illogical. He refuses to accept the possibility faith and hope can lead to a genuinely new and rewarding experience. Spiritual awareness is painful because he has no reason to believe it exists within himself. His ego will not allow him to lean on something within himself, which he knows is irrational. "The only secure truth modern men have is that which they create and dramatize; to live is to play at the meaning of life."[15] Gilbert Chesterton.

Spiritual consciousness expands awe and terror beyond our self to the mysteries of our existence where they belong. Self-worth is no longer about pleasing people. Our worth derives directly from the very source of existence, the mind of God. The opinions of others no longer limit us. We answer to the highest judge, the Creator. If we strive to follow his standards, our conscious is free of all guilt and fear. Anxiety, depression, and despair disappear. We can trust the most significant power of all when we discover how much God loves and cares for us. "If God is for us, who can be against us?" Romans 8:28. With God in our lives, we are truly ready to be the master of our life and destiny. His principles are genuinely appropriate to lead us beyond ourselves. We are free to make our own decisions because God is the guide of our conscience. With God as the power which justifies our existence, we need not fear any man. We are free to be independent and self-assured. There is no greater security than knowing the terrifying power of all cares for us. The fact God is invisible and mysterious is all the better. He has given us the liberty to seek him out on our terms. Kierkegaard would ask, "What do you believe and what purpose does it serve?"

The end of life is one moment in time when no amount of reason can save you. Kierkegaard invites you to consider a scenario you won't soon forget.

You are alone, locked in a room on a sinking ship; Through the porthole, you see pandemonium on the deck as passengers fight for few life jackets and seats on a life boat.

As the deck finally clears, your room is filling with water. No amount of reason can save you. All you can do is cry out to God to save you. Will God save you? It's just a matter of faith.

Conscious Mind

"We do not ask for what useful purpose the birds do sing, for the song is their pleasure since they were created for singing. Similarly, we ought not to ask why the human mind troubles to fathom the secrets of the heavens.

The diversity of the phenomena of nature is so great, and the treasures are hidden in the heavens so rich, precisely so that the human mind shall never be lacking in fresh nourishment." Johannes Kepler.

Descartes declared, "I think therefore I exist." 300 years later, Sartre recognized that thinking is the continuously activated part of the conscious mind, which is guided by the ego. Its needs are always changing. Existence, on the other hand, is a state of being. Sartre rightly concluded the consciousness that knows I exist is not the consciousness that thinks I exist. Descartes statement should be, "I know I think; therefore, I know I exist."

The human mind is the most complicated system known to man. Neurologist Antonio Damasio [47, 48] studies how the brain connects to the conscious mind. The brain has 100 billion neurons with a quadrillion synapses, forming multilayered networks, which create mind symbols from physical objects: images, sounds, touch, smell, taste, pleasure, and pain. Higher level networks integrate these mind symbols into a single experience. Mind symbols are stored in memory to aid in current mental activity. The physical brain is the "wetware," but the mind is the "software."

Primary consciousness lets us know we are alive and connected to the world. Psychologist and spiritual teacher Deepak Chopra wrote., "Consciousness is like wallpaper; everyone can see it, but no one takes much notice of it." The awareness that we are thinking is a deeper level of consciousness. It is the self-aware owner of the mind. It gives us the feeling that our lives have meaning. Eckhart Tolle wrote, "The most significant thing that can happen to a human being is the separation process of thinking and awareness." When we talk about ourselves, the "I" we refer to is not our real conscious being; it is only the ego. When are you mad at yourself, who is angry at whom? Ego can be satisfied while consciousness feels something is amiss. We empower the delusions of the ego by acquiring knowledge, exercising reason, and free will. Free will, the prerogative of the ego, makes us both masters and victims of our destiny. More than anything else, the choices we make reveal who we really are. Conscious awareness is our moral compass, telling us what we ought to do and say, as opposed to what the ego is telling us. Life loses its real meaning when the ego ignores consciousness. Ego suppresses deeper awareness, which underlies all religions, both Eastern and Western.

Morality is the act of consulting with the conscious concerning the thoughts and actions of the ego. A conflict arises within the mind when the conscious tries to interfere with the desires of the ego. Naturalism is a social theory, which gives evolution credit for all of human morality. But morality can't be entirely explained by evolution or social norms because it seldom supports the selfish interests of the ego. The highest moral and ethical values have the status of God's rules. Referred to as objective morality, they are self-evident truths that attempt to elevate humanity above its creature nature.

Neurologist and atheist, Sam Harris, argues objective moral values can be known through reason without resorting to God. This claim runs counter to obvious facts. Hume explains, "Morals excite passions, and produce or prevent actions. Reason itself is utterly impotent in this particular. The rules of morality, therefore, are not conclusions of our reason."

Morality would not even be possible if God had not given us a conscious mind to know right from wrong. Thus, God is necessary to create the conditions for the objective moral values, which separate us from the animals. In actuality, self-awareness has a negative evolutionary benefit because it slows down reaction time.

Science can study the effects of biological and sociological conditioning, but it cannot tell us how we should conduct our lives as taught in Scripture. Harris defines moral value as "the wellbeing of conscious creatures." This begs the question; can immoral behavior contribute to the well-being of those who deny the value of morality? Harris' moral logic is incoherent. Schrödinger wrote, "For a solitary animal egoism is a virtue that tends to preserve and improve the species: In any community, it becomes a destructive vice."

Humans are superior to other animals because of our unique intellectual capacity for free will, which is ultimately the downfall of man. Our ego makes us both masters and victims of our destiny. Psychiatrist Jeffrey Schwartz stated, "Materialism is not true. Mind is a non-material force, which causes changes in the brain." Thoughts are controlled by the ego, which is the legacy of our evolutionary nature, the will to survive and thrive. It reacts to every situation according to its self-interest. It constructs its purpose for living. The ego ignores the authentic self, which is the inner voice of consciousness.

"The essence of all human beings is consciousness." Eckhart Tolle Consciousness is our only connection to reality. It shuts down its connection to the body at night because it knows we need physical rest while it subconsciously keeps on working. While dead to the present, the mind creates its version of reality. Only consciousness separates dreams from reality. "All that we see or seem is but a dream within a dream." Edgar Allen Poe, author and poet.

Consciousness is the biggest mystery in science. It is the self-aware observer who is the gatekeeper to our connection to reality. Consciousness actually cannot even be defined, let alone physically located. Science cannot penetrate what people genuinely feel as expressed by the humanities, arts, psychology, and religion. Mind is a holistic system that cannot be understood by reductive scientific methods. Turning inward into our consciousness is where we find the hidden dimensions of reality.

No one has a clue how something as immaterial as consciousness could arise from something as dumb as dirt. Consciousness has no known evolutionary value. Consciousness has never been a subject of scientific study until now because it is too subjective to be analyzed by scientific methods. Biologist Gerald Edelman, 1972 Nobel Laurette, delved into a deeper consciousness, which he calls "the ability to be conscious of being conscious." It is no wonder science can't find any meaning or purpose in life. Philosopher Daniel Dennett points out that consciousness is like money in that having the thing requires having the concept of it, so he suggests consciousness only emerges when its idea does. [49]

While neuroscience has done much to illuminate the functioning of the brain, much of subjective experience remains a mystery. Some neuroscientists think consciousness is just a component of neural activity. They believe all the secrets of humans are contained within the reductive carbon-based neural networks of the brain. The methods of cellular neuroscience and molecular biology have so far been inadequate to the task of investigating the deepest levels of the human mind. Quantum physics has demonstrated that there are phenomena which engage the mind but are not rational and may confound the study of the mind.

Consciousness may very well be an ethereal quantum field phenomenon. Chopra wrote, "Turning inward into our consciousness is where we find the hidden dimensions of reality. The mystery of reality leads to here, to your inner reality, for if anything is to be real outside us, it must enter human awareness." We already know the conscious mind interacts with quantum events in a way that defies explanation. If string theory is correct, every quantum particle in the cosmos contains an extra seven invisible dimensions, which have been conjectured to be the next great frontier to solving the mysteries of existence. The problem is that physical sciences can only study phenomena that have observable properties of mass or energy. Planck stated, "I regard consciousness as fundamental. We cannot get behind consciousness."

Quantum physics has found the human mind has properties that are unique and distinct from the physical brain. The senses stimulate the conscious mind to activate the perception of matter from nothing more than the presence of elusive quantum waves. This is our best understanding of how the mind acquires knowledge through the conscious act of observation. Consciousness cannot be located in the brain, but the objects it creates are real enough to create reality as it is experienced.

Since there is only one reality, there can be one conscious mind. The collective human conscious mind is intimately connected to the universal mind of God. It is just an illusion of the material world that divides us. As far as we know, the human mind is so unique and mysterious, it can never be fully duplicated artificially in a machine.

Universal Consciousness

"There is a third stage of religious experience, which belongs to all religions, even though it is rarely found in a pure form. I shall call it a cosmic religious feeling. It is very difficult to elucidate this feeling to anyone who is entirely without it, especially as there is no anthropomorphic connection of God corresponding to it." Einstein.[50]

"Quantum physics reveals a basic oneness of the universe." Schrödinger Quantum physics has found the human mind has properties that are unique and distinct from the physical brain. Quantum physics has found every atomic particle has the property of consciousness or free will. The vital force of life is the holistic essence of innate primary consciousness. Quantum fields connect all things into a holistic union that emerged in unison from the creation event. The senses stimulate the conscious mind to activate the perception of matter from nothing more than the presence of quantum waves. Consciousness makes us aware our toast is burning. This is our best understanding of how the mind acquires knowledge through the conscious act of observation. Consciousness cannot be located in the brain, but the objects it creates are symbols that are real enough to create reality as it is experienced. Since there is only one reality, there can only be one Universal Conscious Mind. Consciousness is a singular word which has no plural. The collective human conscious mind is intimately connected to the Universal Mind of God. It is only an illusion of the material world which divides us. Schrödinger said, "To divide or multiply something is meaningless. In truth, there is only one Mind."

Greek philosopher Anaxagoras first used the term Universal Mind in the fifth century BCE. Buddhist philosophy teaches the only reality is Universal Mind. Buddhist monk Cau Chin wrote, "Universal Mind is something to which nothing can be attributed. Being absolute, it is beyond attributes. If for example, it was to be described as infinite, it would exclude from it whatever is finite." This statement asserts humans must have an eternal mind to apprehend the Universal Mind. Muslim poet Rumi wrote, "You are not a drop in the ocean; you are the entire ocean in a drop." This is like saying you are made in the image of the eternal God. The idea Charles Hansel stated, "The Universal Mind, being infinite and omnipotent, has unlimited resources at its command, and when we remember it is also omnipresent, we cannot escape the conclusion we must be an expression or manifestation of Universal Mind. A recognition and understanding of the resources of the subconscious mind will indicate the only difference between the subconscious and the Universal is one of degree. They differ only as a drop of water differs from the ocean. They are the same in kind and quality; the difference is one of degree only."

Consciousness is the highest order emergent property of our mind, which, in turn, is the highest emergent property of the cosmos. Physicist John Wheeler said, "The cosmos could not exist without an Observer, which implies consciousness always existed." The ultimate Observer must be God.

British physicist and astronomer Sir James Jean were asked, "Do you believe life on this planet is the result of some accident, or do you believe it is a part of some great scheme?" He replied, "I incline to the idealistic theory that consciousness is fundamental, and the material universe is derivative from consciousness, not consciousness from the material universe. In general, the universe seems to me to be nearer to a great thought than to a great machine. It may well be, it seems to me, each consciousness ought to be compared to a brain cell in a Universal Mind. What remains is, in any case, very different from the forbidding materialism of the Victorian scientist. The material universe is proven to consist of little more than constructs of our minds. Modern physics has moved in the direction of philosophical idealism. Mind and matter are found to be ingredients of one single system."

Scientific experiments are underway to study what is called a sixth sense connecting human minds. Founder of the "Global Consciousness Project" at Princeton, Roger Nelson, claims his 12-year study proves with one in a billion odds "morphic fields" connect human minds subconsciously. Physicist Eugene Binder states, "Consciousness determines existence. Cosmic consciousness may resolve quantum mechanics." Darryl Dunn, professor of psychology at Cornell, thinks he has anecdotal evidence human consciousness can connect to the future and the past. Quantum entanglement is an ethereal connection that exists outside the spacetime framework. It is the only non-locality phenomenon known to science. Since all matter emanated from the big bang, quantum entanglement suggests there is an unknown force of universal consciousness connecting the human mind with the Universal Mind, which is separate from the physical limitations of the material world.

The human spirit embodies a deep reverence for the sanctity of life and nature that transcends any evolutionary survival benefit. Higher consciousness emerged as a search for meaning and purpose. Religion and culture developed in response to the needs of consciousness. Deepak Chopra said, "Spiritual consciousness is awareness of the universal consciousness of God. High consciousness includes awareness of other people's feelings, respect for all flora and fauna, a cosmic feeling of there is something greater than yourself." Scientists at Yale and Columbia have identified the parietal cortex as the location for spiritual experiences in the brain. "Spiritual experiences are robust states that may have profound impacts on people's lives," said Marc Potenza, professor of psychiatry at Yale. The awareness of a higher consciousness than oneself is not limited to religious experience. He said it is also a feeling of oneness in nature or the absence of self during concerts, sporting or political events, or charitable work. Researchers like Potenza are studying the benefits of spirituality to mental health and medical practice.

5/12/2011, London: "Religion comes naturally, even instinctively, to human beings, a massive new study of cultures all around the world suggests. We tend to see purpose in the world." Oxford University professor Roger Trigg said. "The Oxford study, known as the Cognition, Religion and Theology Project, strongly implies religion will not wither away." Neuroscientist Andrew Newberg has studied this phenomenon in great depth and found supernatural beliefs fit naturally into the mechanisms of the human brain. For example, he found belief in God is common in early childhood, whether or not the parents hold such beliefs.

When we see things as being separate and discrete, this is an illusion. Thus, seeing ourselves as discrete and separate objects is an illusion of our limited senses and the limited perception within our minds. In reality, all matter are wave structures of the universe, vibrating and resonating with everything in spacetime. Einstein put it in a meaningful context, "The true value of a human being is determined primarily by the measure and the sense in which he has obtained liberation from the self. We shall require a substantially new manner of thinking if humanity is to survive." [51] We are all bound together to the earth. The moon is attached to the earth. The earth is bound to the sun, and light connects us to the stars. There is no such thing as empty space. From this knowledge of necessary connections, we can then solve the problems of science, physics, logic, mathematics, biology, psychology, philosophy, metaphysics, and theology. We are privileged to be a part of the Universal Consciousness of God because we are imago Dei.

Today's technology is bringing humanity ever closer to a global consciousness of shared mind. Objective moral imperatives give substance to the meaning of universal consciousness. Could global consciousness lead to increasing awareness of Universal Mind? Can a great leader bring peace on earth as prophesied by the book of Revelations?

Soul Consciousness

"The essence of any religion lies solely in answer to the question; why do I exist, and what is my relationship to the universe that surrounds me? It is impossible for there to be a person with no religion (i.e., without any relationship to the world) as it is for there to be a person without a heart. He may not know he has a religion, just as a person may not know he has a heart, but it is no more possible for a person to exist without a religion than without a heart." Leo Tolstoy, Russian author.

As we know, science can only study objects, which have properties of mass or energy, which fall within the observable scale of current scientific instruments. Not only does this limit what we can know about the material world, but science can also make no claims about the spiritual world of the human soul. Soul consciousness is an experience that is difficult to put into words. It is an innate feeling of our humanness that is the antithesis of the selfish ego. The contemporary approach to the mind/body problem does not attack the existence of a soul. The scientific term for the soul is core consciousness or ground state of being. While neuroscience has done much to understand the mechanisms of the brain, much of subjective experience remains a mystery to science.

As we saw, primary consciousness is self-awareness of the thought process as the perception of given situation that conditions it. Consciousness also informs the mind of the unconditioned self, which is the soul or deepest essence of our being. The unconditioned mind is not associated with mind symbols. The soul is usually referred to as the true self.

4/17/2016, National Geographic Channel: "The study of God," by Morgan Freeman: "If you ask who God is, I would say there is a bit of the divine in all of us, God in you; God in me. The God in me is who I am at my core. God in me is the best version of me. The God in me is who I strive to be, who I was meant to be."

It is the awareness all humans are created in the image of God. God is the Universal Soul, which duels in all souls. Getting in touch with the soul can make everything right again because the love of God becomes real. The awareness of God is real. Subliminal quantum waves connect all things to the mind of God. God acts in the world through the soul of humans. In this way, God is in the world without being part of the world. In the same way, the human soul is in the body, yet exists eternally through its connection to the soul of God.

80% of the world's population believe the human soul is eternally connected to or part of a Universal Mind. This belief is based on tradition, revelation, and experience. The current practice of science cannot confirm or deny this claim, although there is a mountain of anecdotal evidence.

The sages of Eastern mysticism and Western spirituality, in particular, Buddha and Jesus, speak to us of human consciousness that transcends thinking and reason. Buddha sought liberation from suffering by following eight simple truths. Christianity faces suffering head-on. Alfred Whitehead commented, "The Buddha gave his doctrine; Christ gave his life." Christ suffered and died to atone for the sin and evil of mankind that brought suffering and death into the world. Eastern mystics learn to achieve connection to their being through transcendental meditation. Prayer, worship, and meditation can help us slow down and enter an inner awareness of the mind at that moment. Slowing down our mind cultivates contemplation, and by contemplation we can turn inward, beginning to have an inner fulness of feelings and self-awareness, and by having inward fulness, we begin to see the obvious deepener meaning of life. Western religions believe the connection to the soul or being is through prayer and worship of God. To know who you are, you are going to need to spend more time with yourself. Value self-awareness. There is no more satisfaction in life than finding your highest purpose.

James Hillman, the founder of archetypal psychology, says the words soul and spirit are often viewed as synonymous, but they can also refer to antagonistic aspects of a person's spiritual being. Hillman associates "spirit with afterlife, cosmic issues, idealistic values and hopes, and universal truths, while placing the soul in the thick of things, in the repressed, in the shadow, in the messes of life, in illness, in the pain and confusion of love, and in the fear of death." [52] He believes religion has tended to the spirit, often at the unfortunate expense of the soul. "To transcend the lowly conditions of the soul is to lose touch with the soul and a spirituality split away from the soul readily falls into extremes of literalism and destructive fanaticism."

The self is usually too preoccupied with daily living to be aware of the needs of the soul. The soul becomes stranded without a purpose, a significant loss, which puts the soul's eternal connection to God at risk. Hillman's concern is with the soul, which he views as the "self-sustaining and imagining substrate upon which consciousness rests and, which makes meaning possible, deepens events into experiences, is communicated in love, and has a religious concern as well as a special relation with death." When we die, our souls will be judged by God. "Fear not those who kill the body but are not able to kill the soul: but rather fear the One who can destroy both soul and body in hell." Matthew 10:28.

Chopra states, "Recognizing God as the Creator and Source of all that exists transcends religions. The visible reality has an invisible reality as its foundation. Faith is recognizing what it is you don't see is what makes what you do see possible. Your consciousness makes it likely to experience the unseen reality.

You don't see your thoughts, your feelings, emotions, desires, instincts, but they are there. You don't understand love, compassion, joy, patience, but it's here. You grow more profound, then you end up with the spiritual domain. This is the Source of our mind, our body, and the cosmos. Our comprehension of God with our finite minds cannot go beyond the finite threshold of the Infinite Mind.

The Christian view of the soul is based upon the teaching of both the Old and New Testament. The Old Testament contains the statement, "Then shall the dust return to the earth as it was and the spirit shall return unto God who gave it." Ecclesiastes 12:7 In the New Testament, the Apostle Paul states, "And so it is written, the first man Adam was made a living soul; the last Adam was made a quickening spirit." 1 Corinthians 15:45 The majority of Christians understand the soul as an ontological reality, distinct from, yet integrally connected with the body. Its characteristics are described in moral, spiritual, and philosophical terms. As stated by Kant, "We cannot prove a priori the immateriality of the soul, but rather only so much all properties and actions of the soul cannot be cognized from materiality."

Eckhart Tolle wrote, "This timeless now, this endless within the self, is your window into eternity. Find your core identity. It's now; now never ends. It never leaves you. Thoughts come and go, emotions come and go, including all of our experiences, but you are not your experiences; you are a timeless being during the change." To see God, we will be transformed into a glorified body, as Jesus was seen after his resurrection. It is like the metamorphism of a lowly caterpillar into a beautiful butterfly. "When the shadows of this life have gone, I'll fly away." Alison Krauss lyrics.

King David said he was a "worm" before Almighty God; for his humility, God loved David more than any other man. You are eternal and transcendent. This view gives us a bridge between God and us. The span of doubt invites every person to cross it.

The first Hebrew letter, Aleph, is the symbol of the soul, meaning the breath of God. The dwelling place of the soul is unknowable Aleph; it projects into existence everything that exists. The Bible is the instruction code for the soul. "You don't have a soul. You are a soul. You have a body." [3] C. S. Lewis.

Faith can bring the mind and heart into awareness of the soul. Faith allows us to find spiritual meaning in every story in the Bible.

CHAPTER 8

HUMAN SPIRIT

Allegory of the Fall

"One fine morning some chemical scum living in a pothole woke up and thought, this is an interesting hole. It fits me so well, it must have been made just for me. As the sun rose and the air heated, the puddle got smaller, but the scum wasn't worried because his world was meant for him. Later in the day he just evaporated into thin air." British humorist and atheist, Douglas Adams.

Adam's metaphor adequately describes our current predicament. We are the children of chemical scum stuck in a cosmic pothole. If the Creator of the pothole were indifferent to his creation, then his existence would make no difference one way or the other. God would be an emperor with no clothes. Thus, it is insufficient to believe there is a God. It must be established God cares for us more than the dinosaurs and chemical scum. The oldest story in the Torah foretells the complex relationship between God and his chosen species. The iconic allegory of Adam and Eve was passed down orally through over 100 generations before it was put into written form about 800 BCE. The story places God as a participant in the drama of human life. A comprehensive analysis of the human condition is presented in a concise and profound insight into life in a pothole. The story is about everyman, denoted by the name Adam. The story establishes the cultural framework for three great religions.

It took 13.8 billion cosmic years for God to create a spectacular cosmos with splendid creatures made in his image. With great mental powers, fully developed conscious mind, free will, immense willpower, and eternal soul, these creatures had unlimited potential to seek their own destiny in an eternal cosmic garden ruled by moral laws and the laws of nature, according to God's plan, for a picture-perfect world. Adam and Eve represent the archetypes of the species God chose to be his soul mates. Primeval human moral innocence is signified by their ignorance concerning the tree of knowledge of good and evil. The talking snake symbolizes the evil nemesis of God, who preys on the souls of the amoral creatures God chose to be his children for eternity. Humans fell from grace because they trusted their animal instincts more than God's will. Eternal life would have to wait. The human creature was not ready to be in God's presence. He would have to prove his worthiness in the crucible of life.

Ancient biblical writers knew humankind could not avoid displeasing God, but they didn't know why. The problem is our evolutionary survival instincts at work. It is as natural as breathing because we automatically consider our own needs first. Selfishness is really what original sin means, by birth, we are drawn to defend ourselves and to feed our wants first, while God's purpose goes against our natural creature behavior because He wants us to rise above our selfish animal instincts and mature into an unselfish better nature. No person can, without experiencing personal suffering and subsequent personal growth, learn to deny himself and consider others first, but we face a supernatural God with a higher purpose for us and that is what He expects. The Adam and Eve story exposes the downfall of humanity as the desire for reason and knowledge, which can subvert trust in God. Trust in God builds character, compassion, humility, confidence, and inner strength.

We have been separated from the true love of God. We are born into sin. The creature must die both figuratively and literally to be in the presence of God. God introduced entropy, which places a time limit on every living thing in the pothole. Death is the only escape. God separated himself from his creation with an impenetrable quantum barrier that stirred the pothole with violent chaos.

Selfish desire for material and carnal gratification, wickedness, unfaithfulness, and disobedience became the root cause of sin and evil.

The soul was left alone to struggle to free the selfish ego from addiction to carnal pleasures. The humans were left to see themselves as nothing more than carnal matter. Humanity fell victim to his creature nature. What we now experience as existence is only a limited view of a veiled reality hidden from us. Our world is just a harbinger of things to come. "For now, we see through a glass, darkly." 1 Corinthians 13:12.

Now God cannot help us. Humans were not prepared to live in God's garden. They are left with only their conscience to be their moral guide. Humans must struggle to make their world a better place. Only then can God look favorably upon his creation. God has done his part. Now it is our turn to show our love for God and each other by striving to be the best we can be.

The stage is set for the drama of modern human life. Without God in the story, we must revert to the meaningless existence of scum trapped in a dying pothole. The Old Testament chronicles the path of God's chosen people. But the Creator and the eternal nature of the human spirit are common to all major religions.

Parable of Job

A fictional character, professor Jastrow, created by Herman Wouk, an Orthodox Jew, lectures on the meaning of Job's parable.

"God must answer for everything that happens, both good and evil. Who will admit there is no God, that the universe makes no sense?

Job the Jew is a handful of dirt, a worm before God. If God is Almighty, the universe must make sense. But Job did no wrong. If God botched Job's life, He botched creation. Job, a worm, cannot know the magnificence of the heavens. Job is humbled because he knows God's purpose is beyond his understanding.

Job lives 140 years after his tale and prospers more than ever. What about his ten children who perished? Did Job behave better than God? God gave in to Satan, the adversary. Satan needs God's permission to do anything. Let him search his deeds, confess and repent. The missing piece is only what his offense was. Job fights back. The missing piece must be with God, not with him. He is as religious as they are. He knows that the Almighty exists, that the universe must make sense. But he, poor, bereft, boil-covered skeleton, knows now that it does not in fact always make sense, that there is no guarantee of good fortune for good behavior; that crazy injustice is part of the visible world, and of this life.

The whole universe is a botch, and He is not an Almighty God. That Job will never concede. He wants an answer. He gets an answer! Oh, what an answer! An answer that answers nothing. God Himself speaks at last out of a roaring storm' Who are you to call me to account. Can you hope to understand why or how I do anything? Were you there at Creation? Can you comprehend the marvels of the stars, the animals, the infinite wonders of existence? You a 'stinking' worm that lives a few moments and dies.

My friends, Job has won! Do you understand? God, with all His roaring, has conceded Job's main point, that the missing piece is with God. God claims only that His reason is beyond Job. That, Job is perfectly willing to admit. With the main point settled, Job humbles himself, is more than satisfied, falls on his face.

So, the drama ends. God rebukes the comforters for speaking falsely of Him, and praises Job for holding to the truth. He restores Job's wealth. Job has seven more sons and three more daughters. He lives 140 more years, sees grandchildren and great-grandchildren, and dies old, prosperous, revered.

The rich flow of literary Yiddish halts. Jastrow goes back to the lectern, pulls the notes from his pocket, and turns over several sheets. He peers out at his audience. Satisfied? A happy ending, yes? Much more Jewish than the absurd and tragic Iliad.

Are you so sure? My dear Jewish friends, what about the ten children who died? Where was God's justice to them? And what about the father, the mother? Can those scars on Job's heart heal, even in 140 years? That is not the worst of it. Think! What was the missing piece that was too much for Job to understand? We understand it, and are we so very clever? Satan simply sneered God into ordering the senseless ordeal. No wonder God roars out of a storm to silence Job! Isn't He ashamed of Himself before his own creature? Hasn't Job behaved better than God?

Jastrow shrugs, spreads his hands, and his face relaxes in a wistful little smile that makes Natalie smiled too."

In his book "Fahrenheit 9/11," Michael More has the following quote by George Orwell, "It is not a matter of whether a war is real, or if it is, victory is not possible. War is not meant to be won; it is meant to be continuous." The Bible tells us good and evil lurk in the hearts of all men. Satan is real, and he is always present. Satan preys on the hearts of men. Only our better angels have kept us from annihilating ourselves. Satan is doing his work, and the end is ever closer. If we keep God before us, He will be there to help us prevail against evil.

Grimm Reaper

"A person who fears death is not free. Free at last, thank God Almighty I'm free at last!" Martin Luther King Jr., 4/3/1968. He was assassinated the next day in Memphis, TN.

Lying awake in bed at night, the darkness and nightmares whisper about your extinction. Imagine being alone, sealed in a cold, dark coffin. Awake to an alarm, put death away, get on with the business of living, put all your money on red, live as if you are immortal.

When your turn comes, you won't need another dollar, what you will wish for is more time, more picnics. The irony of life is if you are very fortunate, if life has offered pleasure, beauty, and love, along with the pain, then you can only love life. You cower in your dialogue with mortality for a little more time, another round of seasons, grandchildren, great-grandchildren. But the clock is ticking — no more whistling past the graveyard. You watch yourself age, your health declines; you wonder, you lose memory, your appetite is gone, you need full-time care. Your intellect, your beauty, your talent, your wit, your rant, your plans, all of your identity is gone. You meet death naked. A word of praise, a prayer, a tear, and it's over. Alone in cold darkness or embraced by a new light, which is the most fitting end? You made a choice. It's too late to bargain.

Life comes with an expiration date. Nothing is more certain than death. It should be recognized that every society is a substitute or replacement for religion, even if it tries to deny it. Communism, capitalism, atheism, consumerism, humanities, and even science are forms of religion also, though they deliberately avoid religious language. It was a protégé of Freud, Otto Rank, who demonstrated all human culture is sacred.

While fear of death may overcome us on occasion, we would be debilitated if we lived in constant fear. Our instinct is to repress fear. Psychologists have recognized the terror associated with mortality as both practical and universal to all cultures. Repression allows human reason to avoid the confrontation between the contradictions of godlike intellect trapped in a creature of evolution. To primitive humans, life after death was a vague instinct. The pharaohs made elaborate preparation for the afterlife. Socrates wanted to believe in an afterlife as he prepared to die. The Old Testament suggests there is an afterlife but does not elaborate on it. It was Jesus who spoke clearly and forcefully on the subject. His death and resurrection made eternal life the centerpiece of the gospel of Christianity. All Christians celebrate the triumph of Jesus over death on Easter with the praise; "Christ has risen!" All of his disciples willingly died martyrs for their beliefs. For three centuries untold, tens of thousands of his followers fearlessly braved torture and death rather than disavow their faith.

"The sun has one kind of splendor, the moon another and the stars another; and each star differs from all other stars in splendor. So, will it be with the resurrection of the dead? "The body is sown perishable; it is raised imperishable; it is sown in dishonor, it is raised in glory; it is sown in weakness, it is raised in power; it is sown a natural body, it is raised a spiritual body." I Corinthians 15: 41-44.

"I intend to live forever. So far, so good." Steven Wright

Further reading: "The Denial of Death" Pulitzer Prize winner, Ernest Becker. [53] My standing in the field of psychology is strictly due to his book and my hapless experience as a human subject.

After Death

"He knows how we are formed; he remembers we are dust. As for man, his days are like grass; he flourishes like a flower of the field; the wind blows over it, and it is gone, and its place remembers it no more. But, from everlasting to everlasting, the Lord's love is with those who fear him." Psalm 103:14-17.

Hawking stated, "I regard the brain as a computer, which will stop working when its components fail. There is no heaven or afterlife for broken down computers; it is a fairy story for people afraid of the dark." Hawking notwithstanding, the expectation of immortality is nearly universal and not just associated with religion. Alan Seigel showed concepts of an afterlife are universal in all cultures. We are all familiar with the Egyptian fixation on afterlife. Socrates argued the soul is immaterial, leading to metaphysics and transcendence. Ancient peoples, just like us, were attempting to deal with life, death, happiness, God, and the immortal soul. The human body is perishable, but consciousness is immaterial, so it is not necessarily susceptible to destruction as part of the material world. Most people believe in life after death, even though no one can prove it. As a metaphor, we may ask how a caterpillar knows it will become a butterfly? How do millions of monarch butterflies migrate to the same mountainous region of Mexico where they've never been there before and were not even born there? As moral creatures, we all would like to believe there will be justice, joy and peace in an idyllic place we have never been.

Enlightened people don't believe in life after death because science has proven our body disintegrates after death and no evidence of a soul has been found. Our typical experience is not the reality of science. Matter is not at all what it seems and is not the most fundamental nature of existence. Now we have new dimensions of string theory. The solution to esoteric equations, the ethereal Cosmic Code, is the essential ingredient of our cosmos. 95% of the universe is dark energy and dark matter about which very little is known. All generalizations about matter and energy are irrelevant because most of the cosmos is unobservable and perhaps unknowable. Life after death is a real possibility within the knowledge of modern science. There may be nearby dimensions of space we cannot experience within the constraints of our physical framework. There may be other universes entirely different from ours, which we cannot experience. Our cosmos is a spherical object of finite size. It is unreasonable to assume whatever is outside our universe is in any way like our cosmos, or physically infinite. Eternity can now be seen as a coherent concept only because the understanding of reality discovered by modern physics confirms the story in Genesis and the need for some infinite reality.

Elisabeth Kubler-Ross states 59% of Americans say they believe in the afterlife. She claims a dying individual's approach to death has been linked to the amount of meaning and purpose a person has found throughout their lifetime. [54] Out of 160 people in a study with less than three months to live, 157 showed those who felt they understood their purpose in life or found special meaning, faced less fear and despair in the final weeks of their lives than those who had not. In the research of Kubler-Ross and similar studies, spirituality helped dying individuals deal with the depression state more positively than those who were not spiritual.

Bruce Greyson discussed current studies of near-death experiences at the 2008 United Nations conference on the subject. [55] Death is medically defined as no heartbeat, no breathing, and no brain function. Death can occur in many different ways. Lack of blood flow to the brain causes immediate changes to occur, which are damaging to the brain cells due to oxygen deprivation. After a few minutes, these changes become permanent, and the process of cell death begins and lasts for days at a rate of a million cells per minute. The question arises, what happens to the human mind and consciousness during the dying process? At what point in time do they stop? It is not known for sure. It could be a few seconds or a few hours. In the early stages, the damage to the brain is minimal, and the person can be brought back to life, even up to an hour. Because cardiac arrest is synonymous with death, there is a lot of experience reviving patients back from death. After the heart stops, healthy brain function ceases within ten seconds, but there is strong evidence consciousness can persist beyond death if a person can be revived. These findings have led to a clinical study of hundreds of cases of near-death experiences. Generally, patients express calmness, euphoria, a tunnel, bright light, and a panoramic view of their whole life. About 15%, report an out of body experience where they viewed the room from a vantage point near the ceiling and could accurately identify the people in the room and describe precisely what they were doing. The scientific paradox is no reliable and independently verifiable evidence of consciousness, even though the patient is clinically dead and there is no observable activity in the brain. Subtle complexities of the dying phenomenon are now only beginning to be subjected to formal scientific study. Doctors David Kessler [56] and Sam Parnia [57] are conducting scientific studies to understand how it is possible consciousness can persist beyond death.

Many Christians believe just as loving hands greet us when we are born, so will loving arms embrace us when we die. In the tapestry of life and death, we may begin to see connections to the past we missed in life. While death may look like a loss to the living, the last hours of a dying person may be filled not with emptiness, but rather with fullness. David Kessler brings powerful stories from the bedsides of the dying that educate, enlighten and comfort us.

The following verse is the most well-known scripture in the New Testament. "For God so loved the world that he gave his one and only Son, that whosoever believes in him shall not perish but have eternal life." John 3:16. This is a concise statement of God's promise to all of humanity. Jesus gave us unequivocal evidence of the truth of this promise, as was dying on the cross. A murderer on a cross next to Jesus said to him, "Remember me when you come into your kingdom." "Jesus answered him, I tell you the truth today you will be with me in paradise." Matthew 27:42 & 43.

About the ninth hour, Jesus cried with a loud voice, saying, *Elio, Elio, lama sabachthoni?* - that which means, "My God, my God, why have you forsaken me?" Matthew 27:46. This cry is a fulfillment of the same cry of David in Psalm 22:1, one of many parallels between that psalm and the events of the crucifixion. It is difficult to understand in what sense Jesus was "forsaken" by God, his Father. It is certain God approved his work and Jesus was totally innocent, but he had chosen to offer himself as a sacrifice to atone for the sins of mankind. God could not look upon Jesus until this sacrifice was carried out. We are saved solely by the sacrifice of Jesus, which God accepted because of his love for his only Son. In no sense of the word would God have forsaken Jesus except for the moment he sacrificed himself as a lamb unto God. "He took up our infirmities and carried our sorrows, yet we considered him stricken by God, smitten by him and afflicted. But He was pierced for our transgressions." Isaiah 53:4&5.

Since we all face death and its attendant anxiety, why not ease it with the hope of an afterlife. Life is full of sound and fury, but then if there is no hope, life ends in the throes and sadness of abject despair, a tale told by an idiot signifying nothing and soon forgotten. No one other than Jesus has ever claimed to have been able to overcome death. Christianity promises eternal life now, not in the future. If Christ rose from the dead, all facts in the Bible are sufficiently explained. His resurrection is necessary because there is no other plausible explanation for the phenomenon as documented in the Bible. These facts denied the expectations of his followers and the teachings of the Jewish religion. We have no idea what the promises of eternity can deliver. We shall have all the answers to everything we have always wanted to know. No waiting necessary. The biblical view portrays a merger of the present with the afterlife, "the lamb will lie down with the lion." Christ's resurrection has already defeated death for Christians, and believers are members of eternity now. Jesus is with believers now. His victory over death is our victory. Everything we do now is pointing toward heaven. "Then I saw a new heaven and a new earth, for the first heaven and the first earth had passed away, and there was no longer any sea". Revelation 21:1.

When people die, their souls will be judged by God. An eternity in heaven or hell will be waiting. Some Christians hold that the unrighteous souls will be destroyed instead of suffering eternally. Believers will inherit eternal life in heaven and enjoy eternal fellowship with God.

Christians believe babies, including the unborn, and those with cognitive or mental impairments who have died will be received into heaven on the basis of God's grace through the sacrifice of Jesus. [60] The world exists for the body, the body exists for the soul, the soul exists for God.

Jesus Christ, Messiah

"The universe is the stage; Jesus is the play." [62] Malcolm Muggeridge

"Christ is the visible expression of the invisible God. He existed before creation began, for it was through him everything was made, whether spiritual or material, seen or unseen. Through him, and for him, also, were created power and dominion, ownership and authority. Every single thing was created through, and for him. He is both the first principle and the upholding principle of the whole scheme of creation. And now He is the head of the body, the Church, which is composed of all Christian people." Colossians 1:15-18.

A significant thread of the Old Testament is the promise by God of a great leader to come, Messiah, fulfilling over 1,000 years of God centered history. As the most significant prelude to the Christian New Testament, Jesus declared He was the Messiah. Bible scholars have found more than 2,000 specific prophecies of the Old Testament have been fulfilled. Of these 333 prophesies concerning the Messiah were performed by the birth and life of Jesus Christ and the remainder are concerned with his second coming. Professor of mathematics, Peter Stoner calculated odds of one out of 10^{17} of Jesus fulfilling just eight of these prophesies. Such odds are the chances of winning three million-to-one lotteries in a row. No other so-called holy book can come close to such an astonishing record. Only a God who knows the future could have revealed these prophecies to those who He would have recorded them in his Word. Here is what the prophet Isaiah said about the Messiah seven centuries before Jesus was born.

"Who has believed our message and to whom has the arm of the Lord been revealed? He grew up before him like a tender shoot, arid like a root out of dry ground. He had no beauty or majesty to attract us to him, nothing in his appearance that we should desire him. He was despised and rejected by men, a man of sorrows, and familiar with suffering. Like one from whom men hide their faces. He was, and we esteemed him not. He took up our infirmities and carried our sorrows, yet we considered him stricken by God, smitten by him and afflicted. But He was pierced for our transgressions; He was crushed for our iniquities; the punishment that brought us peace was upon him, and by his wounds, we are healed. We all like sheep have gone astray, each of us has turned to his own way, and the Lord has laid on him the iniquity of us all. He was oppressed and afflicted yet He did not open his mouth; He was led like a lamb to the slaughter, and as a sheep before her shearers is silent, so He did not open his mouth. By oppression and judgment, He was taken away. And who can speak of his descendants? He was cut off from the land of the living; for the transgression of my people, He was stricken. He was assigned a grave with the wicked and with the rich in his death, though He had done no violence, nor was any deceit in his mouth. It was the Lord's will to crush him and cause him to suffer, and though the Lord makes his life a guilt offering, He will see his offspring and prolong his days, and the will of the Lord will prosper in his hand. After the suffering of his soul, He will see the light of life and be satisfied; by his knowledge, my righteous servant will justify many, and He will hear their iniquities. Therefore, I will give him a portion among the great, and He will divide the spoils with the strong because He poured out his life unto death, and was numbered with the transgressors. He bore the sin of many and made intercession for the transgressors."

Isaiah 53 The book of Isaiah is so holy it was engraved in bronze, while the rest of the Old Testament was written on parchment. An original was found at Qumran with the Dead Sea scrolls. Not only was it perfectly preserved, but the text is also identical to that in a modern Hebrew Bible.

The following passage from the Book of Daniel was written 200 years before Christ was born. "I saw in the night visions, and, behold, one like the son of man came with the clouds of heaven and came to the Ancient of Days, and was led into his presence. He was given authority, glory, and sovereign power, all people, nations and men of every language worshiped him: His dominion is an everlasting dominion that will not pass away, and his kingdom is one that will never be destroyed." Daniel 7:13, 14.

Messiah, means "anointed one" and Immanuel means "God with us." The Old Testament foretells the coming of the Messiah who will lead the nation of Israel to greatness and power over all their enemies. Many stories from Genesis play a key role in foretelling the arrival of Jesus four millennia later. The religious sacrifice commanded by God of the Israelites was a lamb. God told Abraham, "take your son, your only son, Isaac, whom you love, and go to the region of Moriah. Sacrifice him there as a burnt offering on one of the mountains I will tell you about.?" (now called Temple Mount) Genesis 22:2 God submitted the patriarch of his chosen people to the ultimate test of loyalty, the sacrifice of his son. This incident and the sacrifice of lambs foretold the sacrifice of God's own Son, the Lamb of God, Jesus, the pivotal event of the New Testament. The lamb sacrifice was the highest religious ceremony of the Jewish people. It was practiced throughout the Old Testament. Interestingly, it has not been practiced since the time of Jesus. Neither has Scripture been added to the Hebrew Bible since the time of Jesus. The timespan since Jesus is greater than the timespan of the Old Testament since Abraham. It's time for Jews to accept Jesus as their Messiah. What a joyous day that would be! Christians are adopted Jews because they believe in the God of Abraham. Since the invention of the Gutenberg press, over 65 billion Christian Bibles have been published in 2,000 languages. Today, Christian nations serve as protectors of Israel.

Bono, the lead singer of U2, was asked in an interview, "Christ has his rank among the world's great thinkers. But Son of God. isn't that far-fetched?"

Bono: "No, it's not far-fetched to me. Look, the secular response to the Christ story always goes like this: He was a great prophet. A fascinating guy who had a lot to say along the lines of other great prophets, be they Elijah, Muhammad, Buddha or Confucius. But actually, Christ doesn't allow you that. He doesn't let you off the hook. Christ says, No, I'm not saying I'm a teacher, don't call me teacher. I'm not saying I'm a prophet. I'm saying: I'm the Messiah. I'm saying: I am God incarnate. And people say: No, no, please, be a prophet. A prophet we can take. You're a bit eccentric. We've had John the Baptist eating locusts and wild honey, that we can handle, but don't mention the "M word!" Because you know we'll have to crucify you. And he goes: No, no, I know you're expecting me to come back with an army and set you free from these creeps, but I am the Messiah. At this point, everybody starts staring at their shoes and says: Oh, my God, he's going to keep saying this. So, what you're left with is either Christ is who He said He was—the Messiah—or a complete nutcase. I mean, we're talking nutcase on the level of Charles Manson. I'm not joking here."

Surviving letters of Roman magistrate Pliny, the Younger, document the practices of the early Christians. He described chanting rituals to the deity of Jesus as one of his most significant concerns. He was also concerned because they refused to recognize any Roman gods, even upon the threat of death. One of the chants is called the Kenosis Hymn. It is quoted in an epistle by Apostle Paul to the church in Philippi.

Macedonia, written in 62 CE. The community of believers was grappling with how to serve God best. Paul cited the hymn to illustrate the example of Jesus for them to follow.

The Kenosis Hymn, Philippians 2:5-11.

"Let this mind be in you, which was also in Christ Jesus, who, being in the form of God, did not consider it robbery to be equal with God, but made Himself of no reputation, taking the form of a bondservant, and coming in the likeness of men. And being found in appearance as a man, He humbled Himself and became obedient to the point of death, even the death of the cross. Therefore, God also has highly exalted Him and given Him the name which is above every name, at the name of Jesus every knee should bow, of those in heaven, and those on earth, and those under the earth, and every tongue should confess Jesus Christ is Lord, to the glory of God." The importance of this passage cannot be overstated because it unambiguously declares the divinity of Jesus. The Jewish Christians in Jerusalem were reluctant to accept a belief that conflicted with the monotheism of Judaism. This conflict was not an issue with non-Jewish converts to Christianity. The non-Jewish Gentiles around the Mediterranean came into Christianity from other religions with multiple gods.

The broader importance is a deeper understanding of the person and work of Jesus from the very beginning of Christianity. The Council of Nicaea settled the question in 325 CE where the doctrine of the Trinity, three persons in one Godhead, was enunciated according to the following Scripture: "Jesus came and spoke unto them, saying, all power is given unto me in heaven and earth. Go ye therefore, and teach all nations, baptizing them in the name of the Father, and the Son, and the Holy Spirit, teaching them to observe all things whatsoever I have commanded you and, I am with you always, even unto the end of the world. Amen" Matthew 28:18-20.

Saint Augustine likened the Trinity to a lighted candle; the Father is the candlestick; the light being Jesus and the warmth being the Holy Spirit. Saint Patrick of Ireland likened the Trinity to the Shamrock leaf with three pedals. Since humans are made in the image of God, the Trinity can be compared to the mind, heart, and soul of humans. Life is a journey from mind, to heart, to soul. A human manifest himself as three person he thinks he is, the people other people think he is, and the person God knows he really is. The mind of God is the overarching Father, Jesus represents the love of God, and the Holy Spirit is the very soul of God. As mortal beings, we are temporarily constrained by spacetime, whereas the eternal God is not, except by the choice made by Jesus.

Jesus is presented in the New Testament as a threat to the traditional Jewish and Roman institutions. "The Jews picked up stones to stone him, but Jesus said to them, I have shown you many great miracles from the Father. For which of these do you stone me? We are not stoning you for any of these, replied the Jews, but for blasphemy, because you, a mere man, who claims to be God!" John 10:29, 31. His followers believed He was the Son of God. Jesus is presented himself as a higher authority than the Torah. He is master of the forces of nature; He has power over life and death; He forgives sins and offers eternal life. Only a truly divine entity could demonstrate such amazing supernatural powers. If it is not true, Christianity is the most elaborate and successful conspiracy in history. The New Testament substantiates the importance of Jesus as God's equal. "In the beginning was the Word, and the Word was with God, and the Word was God. He was in the beginning with God. All things came into being through him, and without him not one thing came into being. What has come into being in him was life, and the life was the light of all people." The Word became flesh and made his dwelling among us. John 1:1-5 & 14 The use of Word in the above passage is the original Greek logos, meaning rational thought. The meaning is Jesus speaks for God. Jesus is the Creator's Voice. God's plan was executed by the Words spoken by Jesus. God's plan for the universe is logical and purposeful. Human language is the outward manifestation we are made in the image of God. It is our spoken words, which reveal what we believe and will save or condemn us. The laws of the cosmos are the sensible and rational Word spoken by Jesus, who is the creative source and sustaining power of God's plan for the physical universe. God is revealing his plans to us for our benefit. The Trinity was present at the beginning. The plural word for God in Genesis had to be forgotten because it resulted in confusion. Jesus would have to wait to reveal himself. Even today, the rightful place of Jesus, the Messiah, is rejected by the Jews. Solomon's Temple cannot be rebuilt without the authority of the Messiah. Thousands of years have passed. The time is now.

The incredible and unbelievable act of sacrifice and humility by the Creator of the cosmos for miserable creatures on a speck of a planet makes Christ the biggest story in the history of humanity, bar none. But it is only the prelude to the real story. He voluntarily gave himself as the ultimate sacrifice through humility, pain, and suffering, even unto death on the cross. His dying words were written 1,000 years earlier by King David in

Psalm 22:1. Jesus paid the price for our selfish, prideful sins. He paid the price for the debt we could never pay.

The people of the first century would have considered a bodily resurrection of Jesus to be as impossible as we do today. It was certainly not a practice under Jewish law. The crucifixion meant defeat. Cephas, the high Jewish priest, had the nails from Jesus' cross placed in his Ossuary. Christ's followers denied knowing him and hid in fear for their lives, then three days later, the greatest miracle in human history took place. On the cross, Jesus revealed He was truly a man, for if He was not, we could doubt his divinity.[62] The triumph of the resurrection changed everything. The resurrected Jesus in a glorified body appeared that occupied a place between a visible 4D world and an invisible spiritual dimension. Perhaps a day will come when a great physicist can expand this idea into his extra dimensional string theory.

The New Testament mentions over 100 eyewitnesses saw the glorified body of Jesus during the 40 days after his resurrection. For three centuries, more than 100,000 converts gave their lives for their beliefs. Pascal put it very well, "I believe those witnesses who get their throats cut. It is hard to believe such great self-sacrifice would be done to support a hoax."

The resurrection was so important the Christians changed their day of worship from the Jewish Sabbath to Sunday, the day of the resurrection. Easter is the most sacred Christian celebration. Easter means Jesus conquered death for everyone who believes in him. The life of Jesus is a historical fact, which is better documented than any other person in ancient history.

There is more evidence of the resurrection than any other miracle. If you believe God created the cosmos, then you should have no reason not to think He has the authority to use his supernatural powers to further his purpose. Throughout history, people have always grappled with the resurrection, just as we do today. The Christian transformation gives us the ability to overcome intolerance, violence, injustice, suffering, and death. Without the resurrection of Jesus, Nietzsche could rightfully claim Christianity was for losers. June 5, 2006, Time magazine: Francis Collins vs Richard Dawkins debate

Time: "Dr. Collins, the Resurrection is an essential argument of the Christian faith, but doesn't it, along with the virgin birth and minor miracles, fatally undermine the scientific method, which depends on the constancy of natural laws?

Collins: "If you're willing to answer yes to a God outside of nature, then there's nothing inconsistent with God on rare occasions choosing to invade the natural world in a way that appears miraculous. If God made the natural laws, why could He not violate them when it was a particularly significant moment for him to do so? And if you accept the idea Christ was also divine, which I do, then his Resurrection is not in itself a great logical leap."

TIME: "Doesn't the very notion of miracles throw off science?"

Dawkins: "Francis keeps saying things like 'from the perspective of a believer.' Once you buy into the position of faith, then suddenly you find yourself losing all of your natural skepticism and your scientific – really scientific – credibility. I'm sorry to be so blunt."

Author's comment: Dawkins cannot see the forest for the trees.

We can know Jesus is who He said He was, not simply by his miracles alone, but by the power and authority of his spoken Words, which have been faithfully recorded for ever.

Case for Christianity

"We live by faith, not by sight." 2 Corinthians 5:7.

"Faith is to believe what we do not see, and the reward of this faith is to see what we believe." St. Augustine.

Since ancient times, humanity has sought answers to the philosophical question of existence, but man, by himself, a finite being, could not find the solution. The answer finally came through revelation to the Jewish people. The Words of the All-Powerful God, as revealed in Genesis, tell us God spoke the Laws, which created and sustain the cosmos. According to the discoveries of modern physics, these Laws are carried throughout the universe by quantum waves, which interact to form the illusion of reality we experience. The Creator's Voice, is the source of creation. It's an undeniable scientific theory, if not a scientific fact. In Genesis, God revealed his plan to make creatures who were destined to become his eternal soul-mates. "God had to come to man because man was not ready to go to God." [62] But the creatures ignored him. To fulfill his plan, God entered his creation in the form of a man to deliver his message of Truth and redemption in person. He said; "I and the Father are One." John 10:30. He was crucified for blasphemy. The story does not end here; it is only just beginning; He rose from the dead!

The resurrection is the defining event of Christianity. It is the believe it or not moment that defines the Christian faith. To feel the presence of God through the love of Jesus is a life changing moment. The inspired revelation of Scripture is our only a guide to Truth, not the end of Truth, but the beginning of Truth. The story of Jesus is the greatest ever told. He answers every question of our purpose for existence. The Infinite Personal God has delivered his Truth in person.

His Word is his bond. The Infinite Personal God is our Heavenly Father. He knows our joy, pain, and sorrow because He lived it. He can fill the emptiness we feel inside, comfort us in grief, provide for our needs, give us the courage to overcome our troubles, protect us from evil, give real meaning and purpose to life, and save our souls for eternity. [14]

The writers of the New Testament were sharing the Gospel, meaning "good news," which is "evangelism", not a biography about Jesus. The writers were ordinary men with nothing to gain and everything to lose. The writers were not political nor religious pundits. They were not philosophers creating metaphors and metaphysical spin. They were not historians writing folklore. There is no reason why they were not telling us the truth as it happened. They were compelled by God to reveal the miraculous news to the world. It spread like wildfire. They were ready to die for the Truth. The Spirit of God descended on 500 followers after Jesus had risen, the day of Pentecost. "After his suffering, he showed himself to these men and gave many convincing proofs that he was alive." Acts 1:3. 2,000 years have passed, and no evidence has ever been uncovered to challenge the Truth. All real historical evidence confirms the Truth. There are no contradictions, no need for fabrications, no reason for exaggerations, no incompleteness, nor errors to explain. While eye witness accounts are not always trustworthy, journalist Lee Strobel states the eyewitness accounts of the resurrection have credibility because each version is slightly different. If they were identical, the reports would be under suspicion of conspiracy. [58] It is remarkable the authenticity of the Bible has been preserved. I leave it to experts to interpret obvious symbolism in the Bible. Even scientific explanations for miracles are within the realm of possibility. There is no need for fake evidence. There is no need for underworld turtles, black swans, river sprites, nor red herrings, just a couple of angels to roll away a grave stone.

Now we know why we are here. We were meant to be here by our Creator. The mystery of existence is explained for all time. There is no other possible explanation for why we are here. No other reason makes any sense. Biblical revelation makes good sense because it is the only explanation for everything we know. Philosophy, psychology, and religion are all concerned with the metaphysical problem of being and existence, and Christianity is the answer. Everything we want and need for a good life has been provided for us by our Creator. Everything science has revealed about the cosmos, and human nature is true to Scripture. God is true and faithful to his creation. He gave humanity a mind to know him through his Word and his creation. Through many brilliant minds, science is giving humanity a new revelation of God through his creation. When Christians finally understand the power God has given them, the true knowledge and understanding of God will transform the world. God's plan for humanity is in the Bible, telling us of things to come, even to the end of time. The Bible is the complete answer. God has given enough evidence to satisfy even the most ardent skeptic. Now is the time for all good Christians to embrace the truth of science. Christians have yielded the high ground of reason and knowledge to skeptics for too long. It has been proven time and again nothing can stand against the Truth of God's Word. Nothing even exists except through the power of God's Word.

Christianity is worthy of intellectual discussion and debate. Every field of science is bringing knowledge of God's secrets to the benefit of humanity. To deny God is to become your own worst enemy. Atheism sucks the meaning right out of existence. All atheism offers is a sad dead end. God never intended for death to be the winner. God only wants the best for us.

Why would a cabal of Jewish conspirators conjure up outrageously false stories way after the fact in order to support a religious fanatic who was crucified for blasphemy against their religion? There was nothing for them to gain by spreading fiction and lies, even if the gullible masses believed them. What point is there to a movement whose leader is dead? We must find a way to understand what they believed. It should be clear by now that religion and science harbor a world stranger than fiction. Logic and reason alone are not up to the task of finding truth in a reality that continues to inspire our collective imagination.

Every decent person, including skeptics and atheists, know Christian values resonate with secular morality and ethics to offer us the best way to live with the best outcomes in life, both as individuals and as a society. Atheists are advised to prepare arguments for Judgment Day. But now they cannot claim they did not have enough evidence.

Excerpt from "He is There, and He is Not Silent," by Francis Schaeffer. "We know God is there because He is not silent. His Word and his creation speak to us. He has told us the Truth about himself. He has told us He is the Infinite Personal God. He has given us the answer to existence. He has explained his purpose in simple, unambiguous, and meaningful terms. He told us the faithful would never walk alone. You don't need to be a scientist nor theologian to know the Infinite Personal God. He asks so little of us and gives so much in return. His ways are secure if we trust in his love for us. We ignore him at our peril. Our deepest needs and desires can point us to the reality of God. We want and need something nothing in this world can fulfill. We know there must be more to life than material needs. This unfulfilled longing qualifies as a sincere human desire that is a significant clue that God is there. Selfish greed is a vital clue to our failure to follow Jesus. Eastern religions offer to escape from the world into some transcendental spiritual existence. Christianity is the only religion which provides the hope of eternity through the sacrifice of Jesus.

The living God is here and now. His care and love have been carried by every generation of his followers, even in the darkest of times. Christianity cannot be reduced to modern liberal theology nor narrow-minded conservative Christianity. If we are going to have great, titanic answers, Christianity must be the full biblical answer understood in the whole light of human reason. We need the biblical revelation concerning the Infinite Personal God to know our rightful place in his presence, now and forevermore. Inspiration and revelation must be in accord with modern science. Our physical desires correspond to material needs that satisfy them. We also have a passion for joy, love, and beauty no amount of material things can fill. This is the big hint there is something we want beyond material needs. This unfulfilled longing qualifies as a sincere human desire that is a significant clue God is there." [59] Jesus did not come into the world to tell us what to do. Jesus came into the world to change what we want to do. C. S. Lewis said, "I thought I came to a place; I found out I came to a person."

"All have sinned and fall short of the glory of God." Romans 3:23. We are victim of our evolutionary past. Selfishness, greed, immorality, lust, envy, pride, anger, violence, lying, cheating, theft, fraud, prejudice, hatred, swearing, backstabbing, gossiping, gluttony, sloth: (need I go on) Like it or not, sin is in our genes. We can do nothing to save ourselves. We cannot earn our way; we must find the way. We need a savior who gives us love, forgiveness, hope, and completeness.

"Thomas said to him, Lord, we don't know where you are going, so how can we know the way? Jesus answered, I am the way and the truth and the life. No one comes to the Father except by me. If you really knew me, you would know the Father as well. M From now on, you do know him and have seen him. Philip said, Lord, show us the Father and that will be enough for us. Jesus answered, don't you know me, Philip, even after I have been with you such a long time? Anyone who has seen me has seen the Father. Don't you believe that I am in the Father and the Father is in me. The words I say to you are not just my own. Rather, it is the Father living in me, who is doing the work. Believe me when I say I am in the Father and the Father is in me, or at least believe on the evidence of the miracles themselves. I tell you the truth, anyone who has faith in me will do what I have been doing. He will do even greater things than these, because I am going to the Father. And I will do whatever you ask in my name, so the Son may bring glory to the Father. You may ask me for anything and I will do it." John 14:5-14. Please read John chapter 3.

The privilege of walking with Jesus, your personal spiritual advisor who will never fail you because He is always there. As I look back through life, I experienced miracles that I didn't even recognize at the time. Jesus works in mysterious ways, but I tell you the truth. You will be overwhelmed by his deep love for you. You will even love your enemies. Once you understand what it really means to be a Christian, you will recognize truth that is more powerful than any other spiritual guide. Everything we search for in life is found in Jesus. Jesus is relevant to each person's needs. "And now these three remain: faith, hope and love. But the greatest of these is love." 1 Corinthians 13:13.

CONCLUSION

"To worship God means to recognize mind and intelligence are woven into the fabric of our cosmos in a way that altogether surpasses our comprehension." [1] Freeman Dyson.

"The whole order of things fills me with a sense of anguish, from the gnat to the mysteries of incarnation; all is entirely incomprehensible to me, and particularly my person. Great is my sorrow, without limits. None knows of it, except God in heaven, and He cannot have pity." Søren Kierkegaard.

Truth does not begin with God. It starts with the question of existence. The world is chaotic, transitory, corrupt, and absurd. Can we find truth in such a world? No philosophical nor scientific truth solves the fundamental existential insecurity of human existence. Jesus rejected things of this world to show us how to live according to the facts of a higher reality.

What important purpose does life serve? Why are meaning and values so important? Is the intuition of optimism and hope just vain delusion? To what end do we strive so hard to achieve? Does humanity have no destiny? Could the joy and peace that belief in God brings be missing? As a successful member of an affluent society, do these questions have any relevance? Life's journey is only taken once. Could conflict, uncertainty, and self-doubt be the beginning of faith? Could the optimism of wishful thinking bring the realization of truth where there was none before? Is it sure the throes of death are the endgame? How do we live knowing we are facing our extinction?

What if we could cheat death? There is nothing to lose and all to gain. Can religion stand up to the progress of science? There is more than enough scientific evidence for God even though Joe, the rocket scientist, Ricky Gervais, Stephen Hawking, and Richard Dawkins don't think so.

The intimate connection between the mind of humans and the cosmic order strongly suggests our existence is not the result of some unknowable, pointless cosmic accident. Does our sense of wonder, awe, love, and beauty show us our lives have value and meaning? Is hopelessness all there is, or can we find hope in a broken world? Late Senator John McCain said, "The most marvelous of human achievements is not to lose hope when experience has taught you to hope is for fools." There will always be uncertainty and doubt that cannot be resolved by reason alone. The great people of science and philosophy never denied the possibility of a higher Truth beyond logic and science. They had a vision of a higher, unseen reality that cannot be justified by rationalism alone. Modern physics has shown us the fact that what we experience is contingent on a higher order transcendent reality as revealed in Genesis and understood by Plato.

Neither reason nor faith should be abandoned in the search for the riddle of human existence. Is God justified as the foundation of reality and the solution to the existential human condition?

The Bible does not mince words about the existence of God. Every passage is committed to direct evidence of God, not as a faith, but an absolute fact. There are only two possibilities here. Either God exists, or God is merely an explanation for what we cannot know. You can accept God on faith or as a fact. But if you say there is no God, and you end up wrong, what will be your defense? Do you then blame God for the evils of men? It would be better to think through your issues with God now.

The only way to doubt God rightly and reasonably is to expose other beliefs under each of our doubts and then to ask ourselves what reasons we have for believing it. It would be inconsistent to require more justification for faith in God than not thinking about how you would justify your own belief system. It should be recognized every belief system begins with an element of doubt. Even if you remain skeptical, you will hold your own position with both greater clarity and greater humility.

Scientists are reluctant to say a new theory represents the truth. But that doesn't mean science cannot test theories and find some are far more empirically verifiable than others. A hypothesis is considered empirically verified if it organizes the evidence and explains phenomena better than any known alternative theory. If further testing leads us to expect with accuracy, many and varied events better than any other rival account of the same evidence, then it is accepted theory or even fact, though not in the strong rationalist sense of a rigorous logical proof.

Likewise, belief in God can be considered a hypothesis, which can be tested and verified but not formally proved. The view there is a God leads us to expect the things we observe; a cosmos exists, scientific laws operate within it; it contains human beings with consciousness and free will. How can consciousness arise from inanimate molecules? The hypothesis, there is no God, does not lead us to expect any of these things. It leads us to believe we are the result of unexplainable existence followed by a chain of unexplainable cosmic accidents that create life against all odds, and one such creature attained consciousness of its worthlessness without God.

God offers a credible explanation for all the evidence available to us. God accounts for what we know better than any other alternative account of things. No view of God can be proved, but that does not mean we cannot consider the basis for various religious beliefs and find some or even one, which is the most reasonable. Our hearts hear what cannot be spoken. As pointed out by Thomas Jefferson, "If the God of Scripture exists, critical rationality would be the correct way to approach the question of his existence and purpose." We cannot know the mind of God; all we can know is his Word and what He has done. As in the story of Adam and Eve, when we contemplate our decision concerning God, we are no longer neutral. In the end, the most personal version of our lives will be determined by the choices we make.

Catholic Cardinal Schönborn has rightly dubbed the most fervent of faith challenging scientists as followers of scientism, since they hope science, beyond being a measure, can replace religion as a worldview. It is not an epithet that fits everyone wielding a test tube. But a growing number of the profession is experiencing what one primary researcher calls unprecedented outrage at perceived insults to research and rationality, ranging from the influence of the Christian right on government science policy to the fanatic faith of the 9/11 terrorists, to creationist's ongoing claims. Some are concerned enough to perpetuate the idea publicly, science and religion, far from being complementary responses to the unknown, are at complete odds or, as Yale psychologist Paul Bloom has written bluntly, "Religion and science will always clash."

While a student at Caltech I was well aware God had no place there.

Even so, I have never doubted the faith of my father, but I wasn't sure what it all meant until now. Now I have done my due diligence.

Ants toil away in their worlds, unable to grasp or be concerned with something the size of humans. We work away with little or no concern for God or ants. The ant world is too small and God is too big. We deserve no more interest from God than we have for ants. Our planet is just a blue dot in the cosmos. The real wonder is why the Creator would even notice us. We take 50 million breaths and then it's over. If the age of the universe is scaled down to a lifetime, human life would be over before finishing one breath. Common sense and logic can lead us far astray in matters which extend beyond our daily lives. Modern science has given us good reasons to believe there must be a God. Trusting God is no different from trusting other people. Listen to the song in your heart and see the beauty around you to find the way to everlasting love. The love of life can lead us to the love of God. Love of God is intimately connected to the love of life. Our hearts hear what cannot be spoken. "Love is all you need." Beatles.

"Anyone who does not love does not know God, because God is love." John 4:8–9.

God has declared his love for us. True love is the most liberating loss of freedom of all. One of the principles of respect and trust, whether love of God, romance, family, friends, or country, is we have to lose independence and forego selfish freedoms to attain the greater intimacy of truly unconditional love. Love is the natural source of moral values given to us by God, as taught by Jesus. The greatest of all miracles is real love because it passes all understanding. Dios es Amor. Valla con Dios.

"Finally, it is not a matter of obedience. Finally, it is a matter of love." Sir Thomas More.

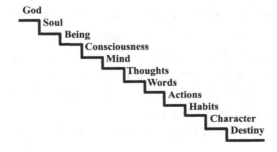

APPENDIX

This is all the math you need to read the book. The large and small scales of physics need to be represented in scientific notation. This is done by using powers of ten, where the exponent equals the number of zeros or digit places in a number as follows:

10^1 = 10 = ten
10^2 = 100 = one followed by two zeros = hundred
10^3 = 1,000 = thousand
10^6 = 1,000,000 = million
10^9 = 1,000,000,000 = billion
10^{12} = 1,000,000,000,000 = trillion
10^{-1} = 1/10 = one-tenth
10^{-3} = 1/1,000 = one-thousandth
10^{-12} = 1/1,000,000,000,000 = one-trillionth

Scale of the universe

Physicists use meters: 1 meter = 39 inches = 3.1 feet. 10 KM = six miles
$10^{10^{120}}$ = number of quanta in the cosmos = Penrose number
10^{80} = approximate number of atoms in the observable cosmos
10^{31} = number of atoms in a typical human
10^{24} = number of planets in the observable cosmos

10^{23} = number of stars in the observable cosmos

10^{20} = number of photons per second emitted by a 100-watt light bulb

10^{18} seconds = 13.8 billion years = age of the cosmos

10^{18} kilometers = distance to Andromeda galaxy (furthest object visible to naked eye)

10^{17} kilometers = diameter of our galaxy

10^{15} = number of galaxies in the cosmos

10^{14} = number of atoms in a human cell

10^{13} kilometers = one light year = six trillion miles (distance light travels in one year)

10^{13} = number of cells in a human

10^{12} = number of galaxies in observable cosmos (97% of galaxies are beyond cosmic horizon)

10^{11} = number of stars in a typical galaxy

10^{10} kilometers = six billion miles = diameter of our solar system

10^{9} kilometers = 186 million miles = diameter of earth's orbit

10^{6} kilometers = diameter of the sun

1.3×10^{4} kilometers = 8,000 miles = diameter of the earth

10^{4} kilometers = 29,000 feet = height of Mount Everest

10^{3} = number of stars visible to the naked eye

10^{-4} meters = smallest object visible to the naked eye

10^{-4} meters = diameter of a human hair

10^{-5} meters = diameter of human cell

10^{-10} meters = diameter of an atom

10^{-15} meters = diameter of atomic nucleus

10^{-35} meters = Planck length (granularity of space)

10^{-43} seconds = Plank time (granularity of time)

REFERENCES

[00] "To copy someone is plagiarism; to copy many is research." Stephen Wright

[0] Biblical references are from New International Version.

[1] "Science & Religion: No Ends in Sight" Freeman Dyson, The New York Review 2002

[2] "The Elegant Universe" Brian Greene, Alfred Knopf 2004

[3] "Mere Christianity" Clive S. Lewis, C. S. Lewis Pte. ltd 1952

[4] From "What I Believe," Forum and Century 84 (1930), 193 to 194; reprinted in Ideas and Opinions, 11

[5] "The God Particle: If the Universe is the Answer, what is the Question?" Leon Lederman, Delta Publishing 1993

[6] Published in 1930 by Science Week, "Glimpses of the Great," "The World as I see it" Albert Einstein, Conf. On Ideas & Opinions" New York 1933

[7] "Reality Is Not What It Seems" Carlo Rovelli, Riverhead Books, NY 2015

[8] From "Physics and Reality", Journal of the Franklin Institute 221, no. three (March 1936), 349 to 382

[9] "Days of Creation; The Science of God" Gerald Schroeder Bantam Books 1994

[10] "The Evolution of Primate General and Cultural Intelligence." K. N. Leland, Philosophical Transactions of the Royal Society, Biological Sciences 2011

[11] Quoted by his Berlin student Esther Salaman, probably around 1920, in Salaman, "A Talk with Einstein," Listener 54 (1955), 370 to 371

[12] "Psychoanalysis and Religion" Erich Fromm 1950

[13] "The Sickness of Death" Søren Kierkegaard, Simon & Schuster 1849

[14] "He Is There and He Is Not Silent" Francis Schaeffer, Tyndale 1976

[15] "Orthodoxy" Gilbert Chesterton, New York 1908

[16] In response to letter from a Colorado banker, August 1927. Albert Einstein Archives Doc. AEA 48-380

[17] Letter to Vero Besso and Bice Rusconi, March 21, 1955. (Albert Einstein Archives doc. AEA - 245)

[18] "The Fabric of the Universe" Brian Greene, Alfred Knopf, 2004

[19] "The Idea of a Personal God" Paul Tillich, Union Review 1940

[20] "Ontological Proof" Kurt Gödel Collected Works: Unpublished Essays, Volume III" Oxford Press 1995

[21] "On the Logic of the Ontological Argument" Paul Oppenheimer & Edward Zalta, The Philosopher's Annual 1991

[22] "The Mind of God" Paul Davies, Simon & Shuster 1992

[23] "The First three Minutes, A Modern View of the Origin of the Universe" Steven Weinberg, Perseus Books 1988

[24] "Astrophysics: The Extravagant Universe" Robert Kirshner, Princeton Press 2002

[25] "Perfect Symmetry, The Search for the Beginning of Time" Heinz Pagels, Simon and Schuster, 1985

[26] In, New York Times Magazine, November 9, 1930

[27] "Thank God for Evolution" Reverend Michael Dowd, Penguin Publishing 1958

[28] "Saving Darwin: How to be a Christian and Believe in Evolution" Karl W. Gilbertson, Harper One 2008

[29] "The Language of Science and Faith: Straight Answers to Genuine Questions" Karl Gilbertson & Francis Collins, Intervarsity Press 22011

[30] "The Language God Talks" Herman Wouk, Hachette Audio 2007

[31] "Genesis and the Big Bang" Gerald Schroeder, Bantam Books 1992

[32] "Quantum Entanglement: Mind into Matter" Fred A. Wolf, Simon & Schuster 1996

[33] "Questions of Truth" John Polkinghorne, Westminster John Knox Press, 2009

[34] "The Intelligent Universe" Sir Fred Hoyle, Oxford University Press 1983

[35] "The Edge of Evolution" Michael Behe, Scribner's Books, 2007

[36] "A Universe from Nothing: Why there is something rather than nothing" Lawrence Krauss and Richard Dawkins, Free Press 2012

[37] "From Eternity to Here: The Quest for a Theory of Time" Sean Carroll, Dutton Penguin Group 2010

[38] "A Grand Design" Stephen Hawking and Leonard Mlowdinow, Bantam Books, 2011

[39] "The Big Picture" Sean Carroll, Penguin House 2016

[40] "The God Delusion" Richard Dawkins, Bantam Press 2006

[41] "The Blind Watchmaker" Richard Dawkins Bantam Press1986

[42] "Denialism, How Irrational Thinking Harms Our Planet and Threatens Our Planet and Threatens Our Lives" Michael Specter, Penguin Press 2009

[43] "When Science Meets Religion" Ian Barbour, Harper San Francisco 2000

[44] "God, Chance and Necessity" Keith Ward, Bloomsbury Publications 2020

[45] "Beyond Psychology" Otto Rank, Dover Books, New York 1948

[46] "The Power of Now" Eckhart Tolle, Namaste Publishing 2005

[47] "Self Comes to Mind: Constructing the Conscious Brain" Antonio Damasio, Pantheon 2011

[48] "Descartes' Error: Emotion, Reason and the Human Brain" Antonio Damasio, Penguin edition 2006

[49] "Brainstorm" Daniel Dennett, Philosophical Essays on Mind and Psychology 1981

[50] In New York Times Magazine, November 9, 1930

[51] Quoted by his Berlin student Esther Salaman, probably around 1920, in Salaman, "A Talk with Einstein," Listener 54 (1955), 370-371, Autumnal Wreath

[52] "The Soul's Code: In Search of Character and Calling" James Hillman, John Lescault, et al. 1997

[53] "The Denial of Death, Ernest Decker, Simon & Shuster, 1973

[54] "On Death and Dying" EliSabeth Kubler-Ross, Google Books 1969

[55] United Nations Conference on Near Death Experiences, Bruce Greyson, Dec. 12, 2008

[56] "Visions, Trips, and Crowded Rooms: who and what You See Before You Die" David Kessler, Hay house Ballantine Books 2010

[57] "Erasing Death" Dr. Sam Partia, Harper Collins Publishing, 2014

[58] "Case for Christ" Lee Strobel, Zondervan 2018

[59] "Life After Death: A History of the Afterlife in Western Religions" Alan Segal, Doubleday 2004

[60] "Life After Death" Dinesh D'Souza, Regnery Publishing 2009

[61] "What's So Great about Christianity" Dinesh D'Souza, Regency Publishing 2007

[62] "Jesus: The Man Who Lives" Malcolm Muggeridge Harper Collins 1975

About the Author

The author received graduate degrees in physics, mathematics, computer science and electronics engineering from Occidental College, California Institute of Technology, and the University of Southern California, where he graduated Magna Cum Laude. He is a member of Phi Beta Kappa and Tau Beta Pi honorary societies.

ON THE COVER

Hubble Space Telescope image of an infant emission-line star located 2,300 light-years from earth. Courtesy of NASA.

Hebrew name of God, YHWH, pronounced YAHWEH, meaning "I AM THAT I AM"

The Hebrew symbol is on the cover:

Printed in the United States
by Baker & Taylor Publisher Services